D0386670

THE SPIRIT OF
MANUFACTURING EXCELLENCE

An Executive's Guide to
the New Mind Set

The DOW JONES-IRWIN/APICS Series in
Production Management

*Supported by the American Production
and Inventory Control Society*

OTHER BOOKS PUBLISHED IN
THE DOW JONES-IRWIN SERIES IN
PRODUCTION MANAGEMENT

Attaining Manufacturing Excellence—**Robert W. Hall**
Bills of Materials—**Hal Mather**
Production Activity Control—**Steven A. Melnyk and Phillip L. Carter**
Use of Microcomputers in Production and Inventory Management—**Thomas Fuller**

THE SPIRIT OF MANUFACTURING EXCELLENCE

An Executive's Guide to the New Mind Set

ERNEST C. HUGE

Principal, Director Manufacturing Excellence
Consulting Practice, Ernst & Whinney

with
ALAN D. ANDERSON

Principal, Manufacturing Consulting
Practice, Ernst & Whinney

Dow Jones-Irwin is a trademark of Dow Jones & Company, Inc.

This book was set in Times Roman by Western Interface, Inc.
The editors were Paula M. Buschman and Joan A. Hopkins.
The production manager was Carma W. Fazio.
The drawings were done by Horvath & Cuthbertson.
Arcata Graphics/Kingsport was the printer and binder.

ISBN 0-87094-989-6

Library of Congress Catalog Card No. 87-70919

Printed in the United States of America

4 5 6 7 8 9 0 K 5 4 3 2 1 0 9 8

This book is dedicated to Gunnar Schmidt,
a most generous and helpful person.

This study seeks to present a succinct but comprehensive perspective on the new world-class manufacturing and its tremendous impact. The book has been honed to a size that today's busy professionals and executives can read in an evening. The book itself is an example of the new philosophy it espouses. All waste has been eliminated, leaving only relevant and high quality material. The many topics covered would otherwise have required a book three times the size. We adopted this approach because it is imperative to see the manufacturing excellence paradigm in the broadest perspective possible. Band-Aid type applications of only a part of the new manufacturing accomplish only a fraction of the potential benefits.

The philosophy of manufacturing excellence arises out of the global competition sparked by the world-class firms of Japan, Germany, and other countries. This new mind-set, probably the most significant advance in manufacturing in the 20th century, is no alien growth; in large part it is the outgrowth of native American ingenuity and industry. The notion that manufacturing philosophies such as just-in-time will not work in the United States has been proved untrue.

The impact of the philosophy of manufacturing excellence is immense. First, the traditional manufacturing function is transformed, and the optimum tactics for automating finally appear. The development of new manufacturing systems such as MRP, OPT, and FMS need to be in delicate balance with the rest of the factory's technology. Second, all firm functions become affected by and integrated with manufacturing. Manufacturing, finance, research and design, and marketing must communicate and together develop strategy.

The new strategy further requires that top management and organizational culture genuinely adopt and act out the spirit and values of manufacturing excellence. This transforming leadership evokes the alchemy of great vision and guarantees the success of the spirit of manufacturing excellence.

ACKNOWLEDGMENTS

We are grateful for the generous help received in writing this book. Reviewers included Jim Burlingame of Twin Disc, Paul Bacigalupo of IBM, Jim Dilworth of the University of Alabama, John Straebel of GENCORP, and Mark Field of Farley Industries.

Special thanks are extended to Ken Stork of Motorola and Bob Hall of Indiana University. Their exceptional commitment to manufacturing excellence is a continual source of inspiration. Their insights have been invaluable.

We deeply appreciate the opportunity to work with such a rich variety of companies. Their confidence in us has deepened our commitment to do more.

We owe a lot to a number of people at Ernst & Whinney, especially to Gerry Vasily for his spiritual inspiration and editorial prowess and to Dick Loehr, for his strong, persistent, and positive support.

Ernest C. Huge
Alan D. Anderson

CONTENTS _____

CHAPTER ONE

Introduction—
The New Manufacturing Paradigm

The advent of world-class manufacturing has clearly ended the age of U.S. industrial supremacy. Beginning about 1965, production in the United States began to experience new and mounting pressures that made much of the art and science of manufacturing obsolete. Soon after, U.S. industries, however unknowingly, began the painful process of replacing the mass production system with a new manufacturing paradigm. The new world-class firms achieved lower cost, higher quality, greater flexibility, and innovative product and process technologies, thus destroying American industry's competitive advantage. Our global competitors revolutionized manufacturing with what legitimately can be considered the most important development in 20th century manufacturing.

The new paradigm is not automation, or robotics, or one of the computer-oriented technologies, although it makes use of these and other new concepts and techniques. The paradigm is what, in lieu of any less accurate but more familiar term (such as Just-In-Time and Total Quality Control), we call the philosophy of manufacturing excellence, or world-class manufacturing. World-class firms, some of them American, scan the globe for the best manufacturing processes and techniques. Thus, they can compete simultaneously on all four competitive dimensions—price, quality, dependability, and flexibility. Their formidable manufacturing organizations characteristically uphold a corporate philosophy of excellence that sustains a meaningful, productive culture and shapes and guides competitive strategy. For world-class companies, the manufacturing function takes on a more equal and proactive role in defining competitive strategy. Taken as a whole, these developments of a strategic philosophy for manufacturing represent the greatest advance in manufacturing since the introduction of scientific management and mass production in the early 1900s. World-class manufacturing provides the best route to the factory of the future.

Two crucial facts arise in all studies of the world's most successful manufacturers. First, the companies that have developed and implemented the most effective manufacturing strategies are those characterized by a strongly held set of values and beliefs, i.e., a corporate philosophy. This philosophy, or spirit, permeates the firm from the top floor to the shop floor. Second, in these world-class firms, manufacturing is integrated into the corporation's competitive strategy; it is consistent with, interacts with, and is supported by the company's competitive corporate rate strategy—manufacturing is viewed as a competitive resource.

To attain world-class quality, a crucial dynamic must exist between the firm's philosophy of excellence, its corporate competitive strategy, and manufacturing. The firm's philosophy guides both the competitive corporate strategy and manufacturing. Moreover, corporate strategy and manufacturing dynamically shape and direct one another, especially as other functions such as marketing, research and development, distribution, and finance come to work more intimately with manufacturing. Manufacturing, through this process, develops a manufacturing strategy, just as marketing or finance have strategies.

These new relationships demand the full involvement and support of top management in manufacturing. Because of the amazing gains world-class manufacturing makes possible, manufacturing comes to be seen as contributing substantially to a firm's competitive advantage. Shorter lead times, lower inventories, faster setups, and quicker delivery responses accompanied by lower cost and higher quality constitute a crucial competitive advantage. These gains give new flexibility and power to marketing, finance, distribution, and design. When a firm develops world-class capacity, all competitive analyses of its industry change radically.

CORPORATE PHILOSOPHY—VISION AND PURPOSE

In the past, tough-minded manufacturers rarely paid attention to such "soft" topics as the philosophy or value system of the organization. World-class manufacturers have known that to survive and achieve success, a sound, empowering philosophy that brings out the great energies of their people is crucial. Faithful adherence to these values is the most important factor in corporate success.

A firm's philosophy is a set of values and principles that guide the firm's goals and strategies. The philosophy establishes what is important to the firm and what is ethically or morally acceptable. The philosophy demonstrates how a firm should treat its personnel, its customers, its suppliers, and the community in which it participates. The organizational culture ("the corporate culture") arises out of the philosophy, permeates the entire enterprise, and endures longer than ephemeral business plans.

A powerful philosophy unites people, gives meaning and purpose to their efforts, and guides decisions throughout an organization. Most critically, world-class firms' philosophy takes precedence over and informs strategic planning. International Business Machines and Toyota are two world-class enterprises that have adopted such philosophies into their organizations. The philosophies uphold such values as respect for the dignity of the individual, excellence in performance, and service.

A corporate philosophy unifies the various businesses and functions of the entire enterprise and limits trade-off decisions on short-term versus long-term objectives. In short, a philosophy helps the firm to accumulate significant competitive advantages that breed success.

The manufacturing function, which usually comprises 80 percent of the firm's personnel and capital, feels the greatest impact from the firm's philosophy. It is imperative that top management and manufacturing managers become transformational leaders instilling an ennobling spirit and culture across the entire manufacturing function.

The spirit of factories has become adversarial and lethargic. Chapter 10 discusses world-class manufacturing's primary goal of the reinvolvement of the factory and creation of a new pride in its work. This is more than half of the struggle facing manufacturing. Once total involvement is achieved, increased productivity and quality, as well as the valuable innovations employee involvement spawns, follow.

THE NEW STRATEGY

The second most important fact about world-class manufacturers is that a manufacturing strategy evolves that is consistent with the firm's competitive strategy. Hence, manufacturing is viewed as providing a crucial competitive advantage to the company. No longer is manufacturing a silent and unequal partner to marketing, R&D, or finance.

The tremendous innovations of the world-class competitors, most frequently Japanese, have laid the groundwork for manufacturing's new role. Their ingenuity and determination have led to strategic advantages through a significant re-creation of their production operations. The philosophy of a world-class enterprise inspired one innovation after another until the art and science of manufacturing, often called mass production, was replaced by the new philosophy of manufacturing excellence. Tremendous reduction in setup time, cycle time, inventories and cost as well as outstanding gains in quality and dependability gave these world-class manufacturers strategic competitive advantage. These countless innovations present these firms with even more process improvements and other opportunities to perfect their industry's potential, benefiting manufacturers, customers, and even competitors. The competitor can embrace these new processes and innovations, but only if the firm also embraces the spirit of an ever perfectible art and science of manufacturing and of doing business, i.e., a spirit of excellence.

Therefore, top management must take responsibility for forging manufacturing into a competitive weapon by formulating a manufacturing strategy focused around the company's philosophy and strategy. No one assumes the production process has been perfected and can be ignored. Manufacturing managers must have a full understanding of the firm's plans and communicate constantly with marketing, R&D, design, distribution, and finance. Ways to allow lower investments, faster development of new products, exceptionally short delivery time, or remarkably lower cost will continue to be discovered. Manufacturing issues can no longer be treated by top management as only operational rather than strategic. Manufacturing's impact on the whole enterprise is great and continues to grow.

World-class manufacturing focuses all decisions and policies in one direction to mutually support one another. It provides a strategy for making consistent and focused policy decisions in the design and management of manufacturing operations. These decisions support both the firm's competitive strategy and philosophy. The world-class manufacturing manager sets his operation's objective based on the strategic needs of the corporation. Decisions about facilities, equipment, personnel, controls, and policies are all harmonized with corporate strategy and philosophy. Manufacturing becomes part of the strategic effort to relate the firm's strengths and resources to market opportunities.

HOW WORLD-CLASS MANUFACTURING BEGAN IN JAPAN

Japan is an outstanding example of how management and employees, by working together, can overcome gross disadvantages and gain world leadership in manufacturing. The Japanese have turned what appeared to be disadvantages into enormous advantages and, in the process, started a manufacturing revolution.

Japan has always faced the problems of very little land or natural resources. Since space was limited for manufacturing plants, the Japanese had to conserve space to an extreme degree, especially if they wanted to expand. Also, they could not afford to waste raw materials on scrapped production, leading them to discover ways to process inventory faster and to increase throughput with near-zero rejection rates in their manufacturing operations. To do this, the Japanese had to improve productivity and quality by the same degree. They learned this from America's own Henry Ford I. Ford, as early as 1910, continually strived for reduced throughput time, realizing it required improved productivity and quality. Ford documented many of the techniques and concepts the Japanese are using today to gain competitive advantage.

After World War II, American manufacturing was the only significant industrial plant left in the world. Because it had no important competition, America became complacent and failed to maintain a strong interest in innovation. Also, the U.S. market was so large it could sustain economic growth by itself. America did not have to think globally to survive. On the other hand, global economic thinking was critical to Japanese economic vitality.

After World War II, not only was Japan's manufacturing base destroyed, but also what it did produce was considered inferior. To help overcome the inferior quality of Japanese products, Kaoru Ishikawa, now a world-famous professor, invited American quality guru W. Edwards Deming to instruct the Japanese on how to manufacture quality products. At that time Ishikawa's father was chairman of

Keinandren, so when Professor Deming asked to speak to top management, they came. In retrospect, this was extremely fortunate for Japan.

Dr. Deming and other American experts to follow, such as Joseph Juran and Armand Fiegenbaum, instructed the Japanese in techniques developed in the United States in the 1930s and 1940s but little used. They preached the techniques to U.S. management, but their message fell on deaf ears. As mentioned earlier, American manufacturers felt no need to implement new techniques to become more competitive because they felt they had no competitors.

The Japanese, thinking the United States was competitive worldwide, believed they would have to do better than the United States in order to compete globally, so the great Japanese companies applied these techniques with near religious fervor. As a result, they now have a competitive edge in quality and productivity (two factors that are directly related) that overcomes by a large margin the deficiencies in resources and geographic distance.

The incredible irony is that Japanese top management applied the concepts and techniques of Ford and Walter A. Shewhart of American Telephone & Telegraph in the areas of quality improvement and statistical quality control. The Japanese probably have the lowest "not invented here" quotient. Many persons mistakenly assume the Japanese propensity to copy what's good means they're not very innovative. Their outstanding process and product improvements over the past 20 years show that this assumption is invalid. The Japanese firms also adapted the marketing and finance functions to the enhanced flexibility of short lead times and smaller inventories. The resulting reductions in cost per unit produced and heightened quality are sources of the high national productivity gains Japan has enjoyed for 20 years.

Waste in all its forms reduces productivity and profits. Today, most U.S. manufacturers are content with a 1 to 2 percent rejection rate, while world-class foreign competition has attained an actual .001 percent rejection rate. Setup/changeover times for machine tools, inventories, and lead times are 5 to 10 times greater in the typical American plant. These and other forms of disguised waste, such as unnecessary movement of materials, poor arrangement of plant equipment, and lengthy inspections, have been eliminated in the world-class Japanese and American plants. In reducing waste, the world-class manufacturer achieves even higher quality; in attaining higher quality, other forms of waste are discovered and eliminated. This process is a spiral of achievement that, after a point, constitutes a strategic advantage.

Other concepts' and techniques' impact goes beyond manufacturing to marketing, distribution, finance, engineering, and other firm functions. Many of these ideas are American concepts by origin and fit with U.S. business traditions and American culture. However, their collective adoption requires a thought revolution by management, staff, and labor.

A certain segment of the manufacturing industry is rising to the challenge of what has been called a "third industrial revolution." Industrialists, adopting the new philosophy of manufacturing excellence and process technologies, have integrated them into their overall corporate strategy. GM's "Buick City" and "Saturn Project" are examples. These initiatives continue to probe for the right balance between high-technology automation and advanced-process strategies. The great Japanese firms already have such balances.

The following chapters seek to clue the busy manager into this important and straightforward world of the new manufacturing paradigm. The self-contained chapters may be read alone or skipped. Chapter 2 presents the essential principles of the new manufacturing philosophy and its key elements of total quality control, employee involvement, and just-in-time. Chapter 3 makes sense out of the deluge of acronyms, the proponents of which claim is "the way" to ensure competitive superiority—JIT, TQC, MRP, KANBAN, OPT, FMS, and CIM. Next, we turn to how the paradigm affects the other firm functions and overall strategy. The manufacturing function must become an equal partner with marketing, R&D, and finance in the strategic planning process. Chapter 4 explores the benefits and implications for cost management and performance measurement.

Chapter 5 provides a framework to use to prioritize and implement the numerous new manufacturing systems, concepts, and tools. As such, it constitutes a road map to the factory of the future in which automation makes economic sense and works. How to avoid the conflicts as well as ensure the winning spirit and innovations of manufacturing excellence is the subject of Chapter 6. Chapter 7 describes top management's role and provides a guide for change. Chapters 8 and 9 describe how to overcome the cultural barriers to change. Finally, Chapter 10 explores the implications of the fact that a firm's people, at all levels, make or break a world-class company. The new manager required to spark the great energies and required commitment is more a transformational leader than a commandeering presence. Such leaders must invoke an alchemy of great vision in their people.

PART ONE

Understanding Manufacturing Excellence

CHAPTER TWO

Defining a New Level
of Manufacturing Excellence

INTRODUCTION

The Toyota Motor Company, using concepts created by Henry Ford I and Walter Shewhart of ATT in the 1920s and 1930s, developed a philosophy of manufacturing excellence that outperforms all past approaches. The great Japanese companies, now so successful in the United States, epitomize the execution of this philosophy. They simultaneously have lower cost, higher quality, better service, and more flexibility than competitors that have not evolved as far with the philosophy. Consequently, the Japanese companies have been immensely successful competing with U.S. companies managed in traditional ways.

This new way of thinking is a philosophical breakthrough, a revolution in thought. Adoption of this philosophy requires a new mind-set, a cultural change within a company. This new culture is not unique to Japan. It builds upon ideas native to the United States. Also, it applies to all manufacturers in process industries, as well as discrete manufacturers, regardless of whether they are made-to-stock, assemble-to-order, made-to-order, or engineer-to-order.

FOUNDATION PRINCIPLES

The philosophy is based upon the two principles of continuous improvement and elimination of waste.

Continuous Improvement

Productivity, quality, customer service, and flexibility in product design and schedule changes must continuously improve. There is no trade-off thinking of quality versus cost. It is possible to improve in all of these dimensions simultaneously. There is always room for further improvement, and one improvement leads to another, establishing a cyclical process.

Elimination of Waste

Waste is something that adds no value to the product. Value is usually added only when raw and/or purchased materials are transformed from their received state into a more finished state that someone wants to buy. For example, a truck manufacturer adds value only by matching parts and assembling them into a finished truck. Everything else is waste. By this definition, waste includes such things as counting parts, all forms of inspection, testing, storage, material handling, generating reports, bad quality (i.e., rework, scrap, warranty, excessive lead times, and inventory).

Inventory is the worst form of waste because it masks problems such as bad quality, machine downtime, availability of tooling, data inaccuracy, poor supply deliveries, and other inefficiencies and chronic problems. Because of its relationship to the problems, Japanese managers have termed inventory "the root of all evil."

PILLARS OF THE PHILOSOPHY

Arising out of the foundational principles are three pillars of tremendous import: employee involvement, doing things right the first time (i.e., total quality control), and extremely fast changeover/setups.

Employee Involvement

Employee involvement (EI) means using the creative energies of all employees to solve problems. It requires a very high degree of commitment to the company by all employees. EI is the most significant dimension the great Japanese companies added to Ford's and Shewhart's concepts. On the surface, EI takes the form of quality control circles (QC circles), which consist of groups of people meet-

FIGURE 2-1
What Is Employee Involvement Worth?

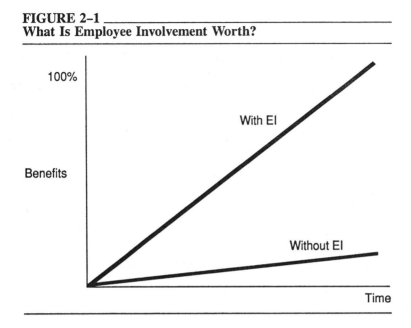

ing regularly to solve problems. Many U.S. companies have tried some form of QC circles, but most have failed because the right culture wasn't in place.

What Is Employee Involvement Worth? Employee involvement is a matter of degree. If EI is implemented to a slight degree (e.g., only 10 percent are involved), significant accomplishments can still be achieved by applying elements of the new mind-set. However, the accomplishments will be enormously greater in an environment where EI is thriving, as depicted in Figure 2-1.

Very few U.S. companies really involve their people yet.

What Does EI Cost? Assuming the right thinking is in place, EI requires a significant investment in time. People must be trained to use problem-solving skills, given time to learn them, and then given time to actually solve problems. Full-time resources are required to institutionalize a problem-solving culture. Facilitation involves providing continued guidance and support. On the average, 5 to 10 percent of budgeted head count should be allocated for this process. Depending upon the leadership, intensity of purpose, and energy allocated to this change, it normally takes companies *at least* five years to institutionalize the process.

Continuous improvement through employee involvement is the credo of the great Japanese firms. The number of ideas and innovations implemented per person are a function of employee involvement. These become important performance measures, as well as invaluable benefits. In 1984, U.S. firms averaged a little less than one suggestion per person. At Toyota the same year, the average was 35 and more than 95 percent were implemented. This is a principal reason why Toyota sets the standard among discrete manufacturers throughout the world. U.S. manufacturers tried to copy the company and created QC circles. But they met with little success because the culture wasn't in place. Most companies didn't understand the culture must be changed. Others have started the long but essential process of cultural change. Chapter 7 discusses how to bring about cultural change.

Doing Things Right the First Time

Traditional manufacturers incur 20 to 40 percent of their total manufacturing cost because of *bad quality,* because things are not done right the first time. The cost of bad quality includes internal failures such as scrap and rework, and external failures such as warranty expense, failure detection/inspection, and failure prevention (e.g., quality engineers establishing quality control procedures).

The approach used to ensure things are made or done right the first time has been termed *total quality control* (TQC). Basic TQC premises follow:

- Quality means satisfying customers' needs, not just internal product specifications that may or may not relate to customers' needs. Furthermore, many customer needs cannot be expressed in terms of quantifiable specifications. For example, automobile purchasers are concerned with feel behind the wheel and the car's general looks. They are not concerned about the actual windshield slant specifications. "Feel" is difficult to quantify.
- Customers are the next operation or person in the total process that ends with the ultimate end user(s). A customer could be the next machine operator or an assembly plant. The customer is "king." Everyone, including marketing, should help with the internal detail of how to better serve the customer.

TQC applies to all operating practices in every function (marketing, engineering, finance, human resources, manufacturing). This ensures customer needs are satisfied by preventing defects and errors.

There are two types of quality:

- *Quality of design.* For example, a Mercedes-Benz 720 has a higher quality of design than the NUMMI Nova. Many quality problems originate in poor design.
- *Quality of conformance.* This is the degree to which design specifications are met.

Traditional U.S. companies ensure product quality (i.e., quality of conformance) by inspection at various stages of the production process, usually by taking samples of various lots. If the sample passes, the entire lot from which the sample is taken passes. This inspection helps to ensure the final product shipped to the customer is not faulty by sorting good lots from bad lots. However, it does little to actually improve quality. Further, as the following chart shows, even though the sample passes, there is a fairly good chance bad parts will get through.

Sample Inspection

Probability that no defects will be found in sample of 10 pieces	
True Defect Rate	*Probability None Found*
20%	11/100
10	35/100
5	60/100
1*	90/100
0.27	97/100

*For 1 percent defect rate: none found 9 out of 10 times. 1% = 4 times higher rate than OEMs permit.

TQC ensures quality of conformance by controlling the variation in the processes used to manufacture the product. For example, a producer of antibodies is concerned about the potency of the active ingredient and the level of contaminants. Key process variables that affect the potency are pH, time, temperature, and the mixing RPM. By controlling these process variables, the quality of the final product is assured.

Some variation exists in every operation of every process. (A process is a sequence of operations.) Relative to controlling the process, the important consideration is whether or not the variation is normal or abnormal. Normal variation is acceptable. It is predictable and random, or doesn't follow a pattern, trend, or cycle. Abnormal variation is unacceptable not only because it causes bad quality, but also because it is an extraordinary occurrence that is due to an assignable cause. The following chart compares normal and abnormal variation.

Examples of Variation

Abnormal (Unacceptable)   *Normal (Acceptable)*

1. Batch of defective raw   1. Random variation in raw
   material   material
2. Faulty machine setup   2. Resin temperature
3. Poor tool design or   3. Injection rate
   machining
4. Uncalibrated test   4. Lack of human perfection
   equipment   in reading instruments
5. Untrained operator
6. Nonrandom variation in
   raw material

Quality of conformance is insured by controlling process variation by the following means:

- Statistical process control (SPC) to identify abnormal variation.
- Problem solving and permanent corrective action to ensure that the assignable causes of abnormal variation do not reappear.
- Immediate feedback from inspection at the source where a part or product is manufactured.
- Fail-safe methods.

Walter Shewhart developed SPC in the 1920s and 1930s. Unfortunately, it never caught on in the United States. Widespread in Japan, it is a principal reason why the great Japanese companies have less than 10 defects per million compared to a 1 to 2 percent (10,000–20,000 per million) defect rate in the United States. SPC uses control charts based upon statistical laws. These laws show that averages of samples follow a normal distribution.

Control limits are set at ± 3S (standard deviation). A standard deviation is a measure of the spread or range of the distribution from

FIGURE 2-2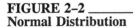
Normal Distribution

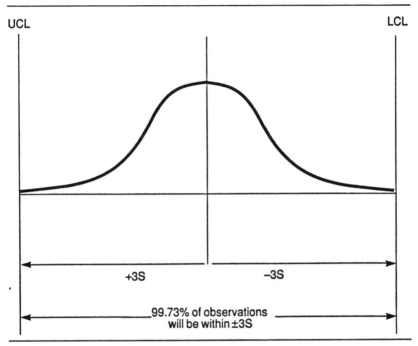

UCL   LCL

+3S   -3S

99.73% of observations
will be within ±3S

the mean. Plus/minus 3S includes 99.7 percent of the total distribution. The components of a control chart are shown in Figure 2-3.

Approach. When a process goes out of control, employees strive to find the root causes. A "cause and effect diagram," as shown in Figure 2-4, can help pinpoint the source of the problem.

To use this approach, the problem-solving team starts with a problem. In this case, it is excessive variation in the test results of a pharmaceutical. By brainstorming, the team identifies all possible root causes. By applying the "80-20 rule," or "Pareto analysis" (renamed after the Italian economist credited with this approach), the potential root causes are identified. To qualify as a root cause, there must be a satisfactory answer to five successive "why questions" about the candidate. After the significant root causes are identified, corrective action is taken. Control charts provide quantitative evidence that the corrective action worked. If it did work, the corrective action is *institutionalized,* that is, implemented permanently. This problem-solving process is summarized by the "Deming Circle"

FIGURE 2-3
Components of a Control Chart

* Center line
* Upper control limit
* Lower control limit

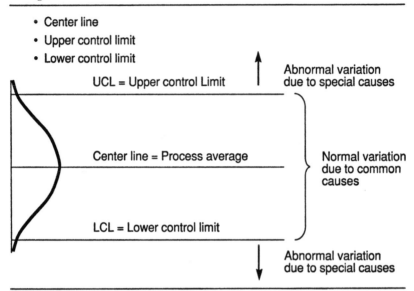

of quality improvement (see Figure 2–5). It was named after W. Edwards Deming, the quality guru who tried unsuccessfully to gain the attention of U.S. managers in the 1950s, but who did get the attention of the now-thriving Japanese companies.

After the abnormal variation is eliminated, the process is said to be in statistical control. When "in control," the process distribution is termed the *process capability.* Even when a process is in control, there still may be a problem meeting specifications if the specification range is less than the process capability as shown:

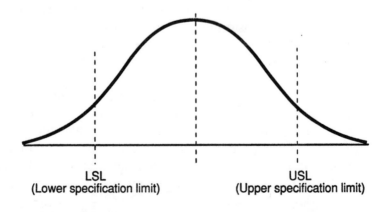

In this case, either the specifications must be changed or the process improved. Ideally, the process capability should relate to the specifications/tolerances, so the capability is well within the spread of the tolerances, as illustrated below.

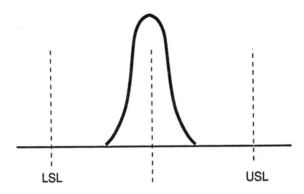

LSL   USL

Once a process is in control, it is predictable and the real work begins to reduce the process variation, which will improve product quality and reduce costs.

Immediate feedback. TQC emphasizes providing immediate feedback to every person whose output does not satisfy requirements. Ideally, the person himself would provide 100 percent inspection. Since testing and inspection is waste, the challenge is to provide 100 percent inspection at the source by spending minimal effort. Direct feedback to the source assumes that if there is a problem, immediate action will be taken if possible. Ideally, the operation will be *stopped* until the problem is corrected.

A fail-safe process. The ultimate way to ensure things are done right the first time is to make the process fail-safe, or build into the operation methods that either:

1. Prevent mistakes or defects, or
2. Instantly detect mistakes or defects.

An example of making the process fail-safe is designing an assembly so it can be assembled only the right way.

By providing immediate feedback, giving operators authority to make corrections, and making operations fail-safe, SPC to control

FIGURE 2-4
Cause and Effect Diagram (fishbone diagram)

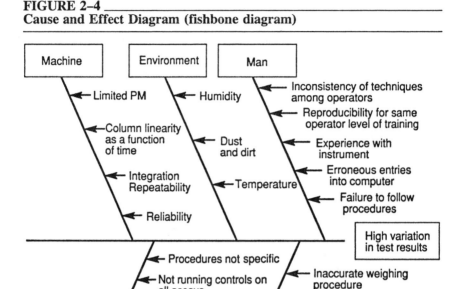

processes is unnecessary. SPC would still be needed to determine process capability and design specification.

Figure 2-6 summarizes and compares the different approaches to managing quality of conformance.

Extremely Fast Changeover/Setups

Traditional manufacturers have done little to reduce setup times. To maximize direct labor productivity, big lots were run so setup would be less frequent. In the new mind-set, setup must be extremely short, lasting no more than 10 minutes. Very short setups allow producing smaller lots, which not only increase flexibility to respond to sched-

FIGURE 2-5
"Deming Circle" Quality Improvement

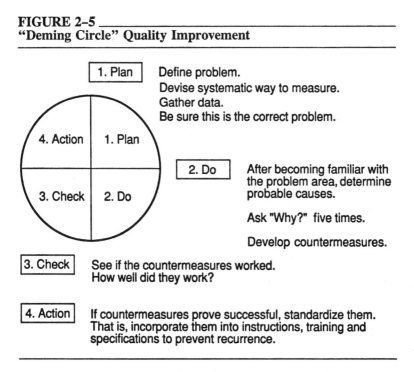

| 1. Plan | Define problem.
Devise systematic way to measure.
Gather data.
Be sure this is the correct problem. |

| 2. Do | After becoming familiar with the problem area, determine probable causes.

Ask "Why?" five times.

Develop countermeasures. |

| 3. Check | See if the countermeasures worked.
How well did they work? |

| 4. Action | If countermeasures prove successful, standardize them. That is, incorporate them into instructions, training and specifications to prevent recurrence. |

FIGURE 2-6
Spectrum of Approaches to Managing Quality of Conformance

Bad	Poor	Good	Very good	Outstanding
Test final product only	• Inspect various stages • AQL for lots	Apply SPC for: • Process control • Process design & capability analysis	• 100% inspection at source • Immediate feedback • Immediate response • SPC for processes design & capability analysis	• Failsafe the processes • SPC for process design & capability analysis

ule changes, but also promote faster feedback of quality problems. Companies that heretofore did not work at reducing setup times are usually able to reduce them by 75 to 90 percent over three to six months, frequently with little, if any, capital expenditures. It sounds too good to be true, but these results are being obtained consistently by manufacturers of all types of products. The following is a breakdown of the approaches and typical results.

Approach	Typical Decrease in Elapsed Time
1. Doing everything possible before the machine must be shut down, such as getting all of the tools, jigs, and gauges close by. This amounts to organizing the workplace—having a place for everything and everything in its place.	40–50%
2. Improving the methods by which the setup is performed. Modifying machines and tools for quick setup. For example, using a hydraulic clamp to hold the piece instead of nuts and bolts.	20–30
3. Eliminating adjustments. Reducing setup times on large presses in the automotive industry from nine hours to three minutes is probably the most publicized example. One aspect of this was standardization of die heights and press shut heights, which reduced the need to make adjustments.	15–20

Setup times are usually reduced so much that even if they are done four times as frequently, the total setup time is much less. Also, streamlining setups eliminates a major source of quality variation. Without even working on quality improvement, reject rates improve proportionally with the reduction in lot sizes.

Most companies start the journey of continuous improvement with setup reduction pilot projects because that can obtain significant results in three to six months. This awakens people to the enormous potential of the philosophy. It is not uncommon to see people who said it could not be done become devotees as a result of these pilot successes.

Employee involvement, doing things right the first time (total quality control), and setup reduction are the pillars of the philosophy. Following are additional elements of the philosophy, all inextricably linked.

Housekeeping/workplace organization. Having everything in its place, clean and ready for use, is extremely beneficial. As mentioned previously, doing this with tools required for setup enables a time reduction of as much as 50 percent. Much time is lost finding what people are supposed to work on or with. In assembly, it consumes as much as 20 percent of the total time.

Cellular manufacturing. All machines required to produce a family of like parts from start to finish are grouped. This contrasts with the traditional method of arranging equipment whereby all similar equipment or production processes are grouped together (e.g., all turning, boring, drilling, milling, grinding, heating, and so on). Figure 2-7 compares a traditional layout with a cellular one.

With a traditional layout, the material flow is much longer and will usually vary from lot to lot for a given part. Usually parts must wait in queue for a considerable time before they are machined or processed due to long setup times and unequal processing times for different parts. In fact, queue time can be 90 to 95 percent of the total elapsed time. Consequently, there is a relatively large amount of work-in-process inventory.

With a cellular layout, the material flow for a given part does not vary and is much shorter, setup times are small, and each operation takes about the same time. Consequently, queues are much smaller and work-in-process inventory much less.

Throughput time, or the elapsed time from start to finish, for a given part is typically 10 percent or less in cells.

On the surface, it appears a cellular arrangement would require more equipment. Although it may demand more of a given type of equipment, overall it usually requires less. The key is extremely short setup times, which typically can free 20 to 30 percent of existing capacity to be rearranged into cells and still have excess capacity left.

Ideally, cells will be "product focused"; that is, the parts produced in a cell will be used in the same product or family of products. Product-focused cells result in the shortest material flows and allow the simplest form of production control.

Some companies are developing cells but are not retaining product focus to the necessary degree practical. That is, they set up a cell to machine parts used in significantly different products, which complicates the material flow and increases inventory and lead time. Often a complex computer and cost system are needed to keep track

FIGURE 2–7

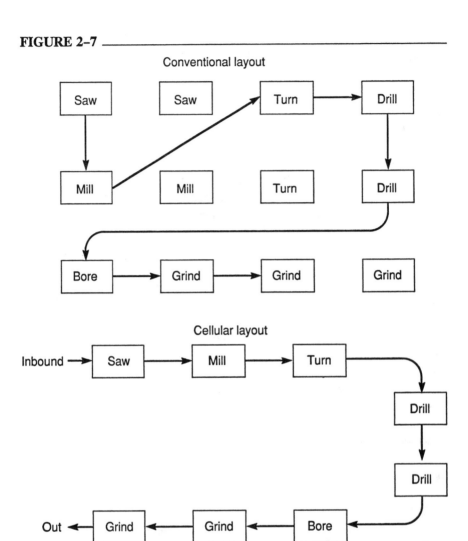

of everything. The ideal is equipment dedicated to producing the product, so material starts at the front end and comes out at the tail end. Volume may not be sufficient to justify dedicated equipment, but if the dedicated equipment is small enough, it may be practical. Unfortunately, many people who select equipment still think in terms of big machines.

Balanced operations. All operations should take about the same time to perform. Without balance, inventory tends to build up around faster operations. Traditional thinkers spend thousands of dollars to reduce the throughput time of a given operation, even though it is already faster than other operations in the total process. This is bad economics. It does no good to go faster than the slowest operation in the total process. Frequently, the existing equipment, although slower than the state of the art, is good enough.

Control by sight. If cells are utilized, spacing compact, and inventory at low levels, the control of material on the shop floor becomes greatly simplified. In repetitive environments, the need for computerized shop floor control systems can be eliminated. This also allows greater worker control of the shop floor.

Total preventive maintenance. Total preventive maintenance (TPM) becomes an extremely rigorous discipline to ensure machines are up more than 99 percent of the time. Ideally, some preventive maintenance is done daily, and routine activities are performed by operators. Traditional manufacturers talk about TPM but few are serious enough to do what is required to obtain 99-plus percent machine uptime. They rationalize that the cost to obtain this degree of running time is prohibitive—that old trade-off thinking again. Enlightened thinkers know that increasing TPM reduces total production costs. In addition to ensuring machine availability, a major objective of TPM is to increase and sustain the improvement of process capability.

Design with productivity, quality, safety, and the environment in mind. Design engineers must design products in a way that facilitates process control. They must assign tolerances with an understanding of process capability, not only of the internal processes, but also of the processes of their suppliers. They must design products with fail-safe assembly.

Ideally, product flexibility will be accomplished by many different combinations of a few standardized modules or parts. When this has been done, product flexibility and low costs have been achieved at the same time. Many traditionalists believed this was impossible. Products should be designed so product options are incorporated by adding parts, at the end of final assembly, to a standard base product.

Customer and supplier networks. Traditional manufacturers have focused on obtaining the lowest *initial* cost by having many suppliers who compete against one another. World-class manufacturers are concerned with the lowest total cost, which includes cost of bad quality and the disruptive cost from parts delivered late. They develop a close, long-term relationship with a few suppliers. Ideally, they treat suppliers as extensions of their own operations. They want their suppliers to adopt the new mind-set. Many manufacturers want them to ultimately demonstrate that their processes are in statistical control, so inspection of incoming parts is unnecessary. The goal is to receive parts and take them directly to where they are used by production. Even if there is more than one good supplier of a part, ideally, only one will be used. In this way, a close relationship can be developed and a major source of material variations eliminated— different vendors have differing process variability. Of course, it will take an enormous time to make this change. Most companies start by cleaning up their own shops before focusing on suppliers. Some require suppliers to give them material on consignment, thinking this is just-in-time manufacturing. Consignment inventory is not JIT. The goal is for the whole logistical pipeline of customers and suppliers to become more integrated and to operate with considerably less inventory.

Pull production. This means nothing is done by one operation until it is signaled by the next operation in the process to do more or make more. If the operation is physically next door, the signal can be, "Hey, Joe, give me more," on an open floor or at a designated location on a workbench. If the next operation is not within sight, the signal can be a card, kanban in Japanese. (Often kanban has been mistakenly meant to mean the total philosophy of manufacturing excellence.)

The pull system of execution requires repetitive production, standard routings (i.e., sequence of operations), and schedules that are leveled sufficiently (i.e., the same or very nearly the same rate every day for a period of two to four weeks). Although the philosophy applies to all types of operations, including job shops, not all operations will be able to have pull production.

Do a little of everything every day. Traditionalists typically manufacture parts in large lots that range from a week's worth to three to six months. After completion, parts are put into the stockroom. An expensive computerized system is required to keep all of the requirements in sync with what is in stock and what is on order. World-class manufacturers do a little of everything every day in final assembly, in parts manufacturing, and at suppliers. This is possible by having extremely short setups, which allow lot sizes of one day for everything. When this can be done, control is greatly simplified and the feedback of quality problems much faster. Doing a little bit of everything every day provides a repetitive pattern that facilitates study for improvement.

The ultimate goal is for production of final assemblies and all other aspects of manufacturing to be in perfect sync with customers' requirements. In this way, inventory will be kept at incredibly low levels by traditional standards. For example, Aisen Warner, a transmission manufacturer, has a just-a-little-more-than-one-day purchased and work-in-process inventory. In applying the philosophy, the manufacturer is unquestionably one of the best in the world.

Results

The results of applying these philosophies are so significant they frequently cause credibility problems. The following have been attained by world-class competitors in Japan and the United States in a wide spectrum of product families.

Implementation Benefits Are Impressive
- 50–100% finished goods inventory reduction
- 70–90% WIP reduction
- 40–70% space reduction
- 30–50% capacity increase
- 70–90% shorter lead times
- 30–50% overhead reduction
- 30–50% product cost reduction
- Rejects reduced from 2% to .001% (10 per million)

CHAPTER THREE

Understanding the Acronym Alphabet Soup—JIT, TQC, MRP, KANBAN, OPT, FMS, and CIM

The past few years have seen a deluge of acronyms. Each acronym represents an innovation, the proponents of which claim is *the* answer. Unfortunately, all have been misrepresented. This chapter defines them and offers guidance as to which are appropriate and when.

JUST–IN–TIME—AN INEXTRICABLE COMPONENT OF MANUFACTURING EXCELLENCE

Just-in-time (JIT) is the core component of manufacturing excellence. It is inextricably intertwined with total quality control (TQC) and employee involvement (EI). Together they constitute the new world-class paradigm in manufacturing.

Unfortunately, the use of the term *just-in-time* has misled managers. Figure 3–1 depicts the different degrees of understanding and how U.S. thinking compares with Japan.

Stage I. Many suppliers think JIT means delivering at a precise time in very small quantities. To do this, suppliers with traditional methods have increased finished goods inventory, which they either store or ship to their customers on consignment. The suppliers believe original equipment manufacturers are using JIT simply to foist inventory back onto them. Similarly, many purchasing people think they have achieved JIT once suppliers start offering them inventory on consignment. Consignment inventory is not JIT. JIT seeks to reduce the overall amount of inventory in the entire chain, from the lowest level suppliers to the end users.

FIGURE 3-1
Spectrum of Understanding about JIT

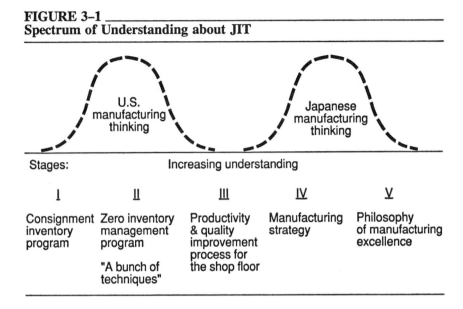

Stages:		Increasing understanding		
I	II	III	IV	V
Consignment inventory program	Zero inventory management program "A bunch of techniques"	Productivity & quality improvement process for the shop floor	Manufacturing strategy	Philosophy of manufacturing excellence

Stage II. Managers view it as a program to reduce inventory, often directed by the materials management organization.

Stage III. JIT is an *ongoing process* (not a program that implies a one-time occurrence) to improve productivity and quality on the shop floor, focusing on improving production processes. Inventory reduction is a by-product of productivity and quality improvements.

Stage IV. JIT is a manufacturing strategy that yields sustained competitive advantage relative to customer service, flexibility, quality, and product cost.

Stage V. JIT constitutes the core of the philosophy of manufacturing excellence. Manufacturing is used in the sense of the *entire* manufacturing company, *not* as the manufacturing function (compared to the marketing, engineering, and finance functions). The philosophy influences and helps form all of these functions. Furthermore, it guides a business on its journey to automation, instead of being in conflict with automation, as many currently believe.

TQC

The critically important concept of total quality control is receiving considerable attention. American quality gurus Deming, Juran, and Fiegenbaum are espousing TQC essentially by itself, which adds to the confusion. Fortunately, Ishikawa, a Japanese quality guru, makes the connection with JIT concepts in his latest book, *What Is Total Quality Control? The Japanese Way.* The philosophy of manufacturing excellence includes TQC as well as JIT. If there is no TQC, there is no JIT. TQC requires that everyone throughout the entire company does what he or she is supposed to do to satisfy customer needs and does it right the first time.

The essence of total quality control includes:

- Knowing customer requirements are the primary imperative.
- Anticipating potential defects and complaints.
- Making all operations fail-safe to the maximum degree, ensuring things are done right the first time.
- Inspecting, when necessary, at the source so there is immediate feedback on problems.

The JIT and TQC philosophy represents a cultural change (i.e., changes in values, in attitudes, and in the way people think). The following are part of this cultural change:

a. Stopping production for quality problems.
b. Giving operators discretion whether or not to stop the line.
c. Recognizing that inventory size reflects the overall productivity of an operation. Large inventories mask problems. Operating with less inventory requires productivity and quality improvements.
d. Manufacturing only what is needed in the immediate time frame; not building inventory just to keep workers occupied and machines utilized; only producing at a rate equal to the rate of the slowest operation in the total process. If a manufacturer produces at a faster rate, inventory will appear in front of the slower operations and inventory is the worst form of waste.

   Producing at the slowest operation's rate has an important influence on technological requirements—frequently, older machines are fast enough. Traditionally, U.S. industry has focused on reducing the machine run time of operations by employing more expensive machine tools. Ironically, this has frequently only increased the plant's break-even point and resulted in building more inventory. The "wiz-bang" machine tools had to be fully utilized so their projected return on investment would be realized (at least on paper).

e. Realizing direct labor efficiency and utilization are invalid measurements. They encourage building more inventory than needed and suboptimizing direct labor productivity at the expense of the productivity of the total process.

MRP

Material requirements planning (MRP) is a computerized manufacturing control system. The MRP process starts when items on bills of material are exploded/multiplied, by a master schedule, to determine material requirements. Next, stock on hand and material on order are subtracted from the exploded requirements to determine what still needs to be ordered. All of these calculations are offset by the appropriate lead times required to produce or procure parts. Although MRP logic is straightforward, it requires a computer to manipulate all of the data. MRP is a "push" system, meaning items are produced at the time required by a predetermined schedule. In contrast, the Japanese kanban approach is a "pull" system.

KANBAN

A pull production system indicates the full attainment of the manufacturing excellence philosophy, although it will not be possible in all operations. In a pull system, parts are produced only if they are needed by and a signal received from subsequent operations. A signal can take many forms. If the "customer" operation is next to the "supplier" operation, the signal can be verbal (i.e., "I need more parts") or an empty designated area. If the operations are separate, the signal can be a *kanban,* which is the Japanese word for *card.* It is essentially an authorization to move or produce parts. Unfortunately, kanban has become synonymous with the total philosophy and further contributed to misunderstanding.

An advanced pull system intentionally produces what is needed in an immediate time frame, such as a day's requirement or perhaps only one part. This is possible only in a repetitive environment where schedules can be leveled for a fixed period and routings are standard.

MRP (push) and kanban (pull) systems are manufacturing control systems that implement the world-class philosophy. Often both will coexist within the same facility. Where a pull system is practical, MRP will still be used to develop the plan and a pull system used to execute the plan. However, as JIT and TQC principles are implemented more extensively, MRP becomes simpler.

MRP nets out inventory and provides a lead time offset. When stock is very small and lead times are very short, MRP becomes a gross requirements explosion (schedule times the bill of materials), with no lead time offset and no subtraction of stock on hand.

OPT

Optimized production technology (OPT) is another manufacturing control philosophy. OPT will help a manufacturing plant driven by traditional measurement shibboleths—direct labor efficiency and utilization, machine utilization, and EOQ lot sizing. Some OPT and manufacturing excellence principles are the same. However, the OPT principles alone will not enable U.S. industry to be world-class competitors. A company following these principles may take a tactical approach of scheduling around existing problems instead of focusing on the root causes of problems such as quality rejects, machine downtime, or excessive setup time. The OPT approach calculates the optimum solution based on current process capabilities and inefficiencies.

By comparison, manufacturing excellence does not accept variation as fixed and strives to eliminate it. A philosophy that attacks and eliminates variation will prevail over one that assumes variations/inefficiencies will continue. In fact, when the variations are as small as they become in plants running according to world-class principles, companies do not need the expensive computerized shop floor control system OPT requires.

FMS

Flexible manufacturing system (FMS) uses production technologies to implement the world-class philosophy. It represents the most highly automated way to implement the JIT concept of cellular manufacturing/group technology. FMS arranges machines to form a cell that machines a group of parts from start to finish. The most widely used definition of FMS is a cell of computer numerically controlled (CNC) machines, with automated material handling between machines. The FMS cell frequently includes a coordinate measuring machine (CMM) to provide automatic inspections on both in-process and finished work. All machine operations, inspection, and movement of material between machines is controlled by a host or central computer. Once material is loaded into the system for the first operation, it is usually untouched by human hands until the last operation. Some manual intervention may be required for tool changing and in-process inspection inappropriate for the CMM. However, in some flexible manufacturing systems no planned manual intervention is required once parts and tools are loaded. The following chart depicts the different degrees to which cells can be automated and where FMS fits in.

Degree of Automating Manufacturing Cells

	I	II	III (FMC*)	IV (FMS)
Machine tools	Conventional	Conventional and NC/CNC	NC/CNC only	CNC only
Material handling	Manual	Manual	Manual	Automated
Overall control of the cell	Manual	Manual	Manual	Computer

*This stage has been termed flexible machining cell (FMC).

It is important to understand that FMS is *a* way, not *the* way, to execute the manufacturing excellence philosophy. The FMS market was projected to explode in the 1980s to meet the onslaught of foreign competition. It has not developed as the machine tool manufacturers predicted because many companies could not afford them. Full-fledged systems cost from $10 million to $20 million. In retrospect, this is fortunate because many benefits of automation are being obtained for substantially less than the cost of building an FMS. This has been accomplished by utilizing and rearranging existing equipment (Stage I) and applying other concepts such as statistical process control, setup reduction, inspection at the source, and preventive maintenance.

Some companies have implemented an FMS only to see the bottom fall out of the market the FMS supplied. They were stuck with an enormous fixed cost. The FMS is flexible only within the product family or geometric envelope for which it is designed. This is especially true when the FMS has automated material handling. Considerable additional investment is frequently required to accommodate new parts outside the product or geometric window.

Some FMS's are not paying for themselves because of traditional machine downtime and quality problems. Since machining and material handling processes are closely linked, the entire FMS, or a large part of it, can go down if one process goes down. To gain sufficient utilization, much better than traditional uptime and quality are required. Many FMS's have been implemented without any system for process control or preventive maintenance—two manufacturing excellence basics.

The FMS is appropriate in certain applications and will be used to a greater degree in the future. However, it should be implemented only after the basics are in place. By employing the manufacturing excellence philosophy, companies not only prepare the way for future automation such as FMS, but they also will generate sufficient cash from improvement to afford it.

FMS's are a subset of the computer integrated manufacturing concept.

CIM

Computer integrated manufacturing (CIM) uses a computer to integrate engineering and manufacturing processes from product inception to product service (see Figure 3–2).

CIM's ultimate vision is 100 percent customization of product produced instantaneously on demand at low cost. Elements of CIM, also referred to as islands of technology, follow.

- *Computer aided design (CAD)* utilizes the computer to design parts and products and eliminates the need for engineering drawings and draftsmen. Pioneered primarily by the aerospace and automotive industries, this technology is expensive and has a long learning curve but is valid in many applications.

FIGURE 3-2

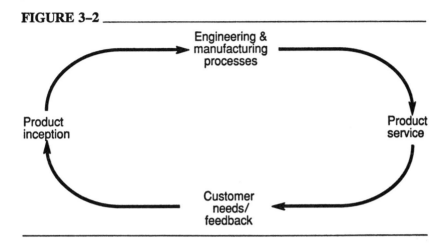

- *Finite element modeling* uses the computer to analyze design integrity (e.g., aircraft strut analysis).
- *Manufacturing simulation* analyzes various plant layouts and machine combinations relative to throughput time and work-in-process inventory levels.
- *Automated process planning* uses the computer to determine the sequence of machining operations and machines used to produce a part from start to finish.
- *Group technology* (GT) groups parts that have common characteristics such as geometry and material. GT has several applications, the most valid relating to product design. When a design requirement is received, the GT system determines if a part similar to the design specifications has already been designed and placed on the computer file. Perhaps the part already designed can be used, thereby eliminating the need for a new part. GT minimizes the proliferation of new part numbers. GT can also group like parts that could be manufactured by the same equipment. Therefore, GT can be used to design manufacturing cells. However, using GT to design cells is seldom justified, especially when cells are focused around products as they should be. Just loading the computer files to be able to use GT can take several years.
- *Distributed numerical control* uses a computer to control all of the machines, which are computer numerically controlled.

- *Flexible machining systems* are similar to distributed numerical control, along with *manufacturing planning and control systems.*

If all the CIM elements were in place, an engineer could, upon receiving a design requirement, scan the GT file to see if a like part had been designed. If not, the engineer would use CAD to design it, perhaps using finite element modeling if the design integrity required it. Then the engineer would use automated process planning to determine the sequence of operations of the various computer numerically controlled machines. Theoretically, an engineer could, after designing the part, develop the programs to control the computerized numerically controlled machines that actually machine the part. Some companies are doing this on a limited scale.

CIM has been used synonymously with "factory of the future" or "factory with a future," implying that if you do not adopt it you will not have a future.

CIM is also *a* way, not *the* way, to execute a philosophy of excellence. CIM will be much easier to implement in a world-class environment. In fact, even if the computer system interface problems arising from integrating all of the CIM elements are solved, CIM will not succeed in factories where traditional thinking persists. Before implementing CIM, the basics must be mastered to ensure waste is not automated. One CAD implementation was based on how many new part numbers would not be needed. But after CAD was implemented, new part numbers proliferated. The computer allowed this to happen. The basic disciplines needed to use this new tool appropriately were missing, and the computer enabled engineers to design parts faster. This example does not invalidate CAD, an extremely powerful technology. It simply illustrates the need for basic disciplines to be in place before implementation.

Summary

JIT should be thought of as the comprehensive and integrative philosophy of excellence in manufacturing enterprises. It embodies total quality control. As such, the philosophy can guide companies relative to different aspects of automation such as FMS. JIT basics (cellular manufacturing/group technology, statistical process control, setup reduction, preventive maintenance, workplace organization, and so on) must be mastered before paying the high cost of automation. Otherwise inefficiencies may be automated or expensive equipment justified by benefits that were, for the most part, obtainable by low-cost application of manufacturing excellence concepts.

Optimized production technology is a philosophy of production control that focuses on the shop floor. Some OPT and JIT principles are the same. However, JIT addresses the "whole world"; OPT does not. JIT focuses on solving root cause problems and OPT on eliminating bottlenecks. The OPT methodology leads to a computerized scheduling process (which OPT sells) to help schedule work through bottlenecks. The scheduling algorithms upon which the computer programs are based are not divulged by OPT.

The MRP manufacturing control system will be needed even when manufacturing excellence principles are implemented to the furthest degree (i.e., pull system), albeit in a much simpler way. In this case, MRP will generate intermediate and long-range visibility of capacity and material requirements.

PART TWO

Strategic Implications

CHAPTER FOUR

Performance Measurements and Financial Control

IMPACT ON PERFORMANCE MEASUREMENTS

For world-class operations, many traditional measurements no longer apply. Direct labor productivity and its components, efficiency and utilization, are an excellent example. They are usually defined as follows:

$$\begin{matrix} \text{Direct labor} \\ \text{efficiency} \end{matrix} = \left(\frac{\text{standard hours earned}}{\text{hours worked}} \right)$$

$$\begin{matrix} \text{Direct labor} \\ \text{utilization} \end{matrix} = \left(\frac{\text{hours worked}}{\text{hours paid}} \right)$$

$$\begin{matrix} \text{Direct labor} \\ \text{productivity} \end{matrix} = \left(\frac{\text{standard hours earned}}{\text{hours paid}} \right) = \begin{matrix} \text{utilization} \times \\ \text{efficiency} \end{matrix}$$

These measurements are inappropriate for the following reasons:

1. They all promote building inventory beyond what is needed in the immediate time frame. Building inventory ahead of need covers up problems and is, therefore, the worst form of waste.

2. Emphasizing performance to standard gives priority to output, at the expense of quality. Relatively few companies even adjust results to reflect bad parts. Furthermore, standards are limiting relative to continuous improvement. Once standards are attained, people usually feel they have "arrived."

3. Direct labor in the majority of manufacturers accounts for only between 2 and 10 percent of total product cost. Focusing further on direct labor is a classic case of overcontrol. Traditionalists have run with very tight direct labor and much looser overhead. Frequently, direct labor head count reductions have been more than offset by overhead increases.

The new mind-set of employee involvement requires that machine operators be given time to solve problems. When there are no immediate requirements for parts, operators do routine preventive maintenance and work to improve the processes. Because employees are the most directly involved with the processes, they know better than anyone what the problems are and what needs to be done. When employees are really involved, overhead is viewed as a resource of line management.

Instead of measuring direct labor productivity, the productivity of all personnel should be calculated. To do this, output should be divided by total head count (everybody on the payroll—administration, overhead, and direct labor). Output should be whatever the operation typically uses to measure output (e.g., gallons, barrels, engines, valves). If there are severe changes in product mix, use manufacturing cost dollars. Trying to be more specific than this is not realistic.

Using machine utilization is similarly inappropriate because it results in building inventory ahead of needs. Focusing on this measurement has frequently resulted in utilizing expensive equipment, and sometimes entire plants, around the clock. Virtually no time is allowed for preventive maintenance; equipment is run flat out until it breaks down, which causes considerable disruption. Furthermore, planning to run on three shifts allows no contingency for changing schedules and reacting to other problems that may arise. Focusing on machine utilization also encourages companies to dispose of older equipment, although still reliable, because newer, faster equipment is available. This old equipment, which can be utilized only a small percentage of the time, might better be dedicated to making a certain product or family of parts. This would simplify the production flow, improve throughput, and reduce inventory. The reduction of inventory may itself exceed the benefit in improved machine utilization overall. In fact, the equipment may already be fully depreciated.

Return on total assets is what counts, not whether equipment is utilized. Total assets includes inventory, which is frequently overlooked when evaluating decisions. Many companies following manufacturing excellence also use return on net assets.

Cost of quality (COQ) or cost of bad quality is being used more often to measure performance. It estimates the cost of not doing things right the first time. COQ includes internal failure costs (i.e., scrap and rework), external failures (warranty, recall, field service, and so on), assessment (inspection), and prevention (quality engineering). When first estimated, it usually ranges between 20 and 30 percent of the total product cost for discrete manufacturers! The measurement quickly attracts management's attention relative to the opportunity improved quality offers. However, once leadership understands COQ and realizes total costs will decrease as quality (reduction of variability) improves, then COQ is a less useful measure. However, it is not counterproductive. World-class manufacturers consider customer feedback, response to customer feedback, and defect rates to be the truest measures of quality.

Schedule Performance

Traditional manufacturers have usually been inward focused and primarily concerned about performance to internal schedules. Enlightened thinking focuses outward on the customer. To them, *customer service* is critical.

Inventory

It is important to relate inventory to time. This can be done by using days on hand, but not turnover, even if generated from the same data. Inventory reflects the overall level of quality and productivity. However, inventory reduction should not be an end in itself. Instead, inventory reduction should be viewed as the result of focusing on improving the process. This is a critical point.

New Measurements

Important new measurements are:

- Ideas generated and the percentage implemented, which reflect the degree of employee involvement.
- Lead time reduction for new product introduction and throughput or production cycle time.

Reduced throughput time simultaneously yields heightened productivity, quality, flexibility, and service. It must dominate operations thinking. Setup reduction is a key to improved throughput and must become a critical subset of performance measurements. Operations must become completely proficient at setup.

Machine availability is critical. Does equipment work when needed?

Traditional emphasis on cost variances (another internal preoccupation) gives way to emphasis on reducing total costs and a constant analysis of the competition's costs. Few companies today know their competition very well. Keen competitive awareness is essential to enlightened thinking.

Traditionally, factory employees' pay was based upon seniority. Instead, for world-class operations it is based upon improved knowledge and capability to do a variety of jobs.

Individual incentives based upon quantities produced are counterproductive because they promote excess inventory. Companies that have had individual incentives but are now implementing the new concepts end the incentive program but pay employees a flat salary based upon the last three months of incentives. Productivity gains due to the new concepts typically reduce average labor cost per unit of output by 10 to 40 percent.

Enlightened measurements relate primarily to the total output of the operation or, if product-focused cells are developed, to groups of products manufactured in the cells.

Group measurements reduce the parochialism and suboptimization of individual incentives achieved at the expense of the whole; they promote cooperation.

After implementing product-focused cells, some companies change from individual incentives to group incentives. If the group incentives are still based only on quantity produced, then they still promote making excess inventory. The company is asking the people to make only what is needed, but is paying them extra to make more.

If a significant percentage (i.e., 50 percent) of total compensation is due to semiannual or annual bonuses, then the bonus can be used to stabilize employment; that is, when business is down, the total bonus is reduced instead of laying off people. If, for example, the bonus pool is reduced by 80 percent, then everyone receives 80 percent less bonus. It is equitable for all concerned.

One significant advantage for great Japanese companies is the employment stability that results from their compensation scheme. One need not look to Japan, however, for the nearly "model" plan.

Created during the Depression, Lincoln Electric Company's profit-sharing/group bonus plan* still serves as a model of fairness and simplicity, compared to the other well-publicized plans, such as Scanlon. Paradoxically, Lincoln also has an individual incentive plan. Unfortunately, some companies that understand individual incentive plans are negative, overlook Lincoln's excellent profit-sharing plan. James Lincoln, chief executive officer at the time, created the profit-sharing concept. A deeply religious man, he based the plan on Christ's gospel. Lincoln Electric's long-standing success speaks well for its strong sense of Christian values. Since its inception during the Depression, the yearly bonus has usually been about 100 percent of the regular compensation.

At the end of the year, the Lincoln board of directors allocates a certain share of earned surplus to a capital expenditures and a contingency "rainy day" pool. The remainder becomes the bonus pool. The bonus funds are divided by all of the year's wages and salaries to determine a factor.

Individual bonuses, including the CEO's, are determined by multiplying this factor times the individual's total earnings times a performance factor. The performance factor, a function of quantity,[1] quality, safety, cooperation, and improvements, is a weighted average

**Lincoln Electric's Profit Sharing Plan Is a Model of
Simplicity and Fairness**

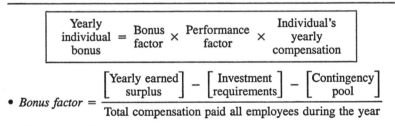

$$\begin{array}{c} \text{Yearly} \\ \text{individual} \\ \text{bonus} \end{array} = \begin{array}{c} \text{Bonus} \\ \text{factor} \end{array} \times \begin{array}{c} \text{Performance} \\ \text{factor} \end{array} \times \begin{array}{c} \text{Individual's} \\ \text{yearly} \\ \text{compensation} \end{array}$$

- *Bonus factor* = $\dfrac{\left[\begin{array}{c}\text{Yearly earned}\\\text{surplus}\end{array}\right] - \left[\begin{array}{c}\text{Investment}\\\text{requirements}\end{array}\right] - \left[\begin{array}{c}\text{Contingency}\\\text{pool}\end{array}\right]}{\text{Total compensation paid all employees during the year}}$

- *Performance factor*
  —Ranges between .8 and 1.2
  —Depends upon quantity, quality, safety, cooperation, and ideas
  —Averages 1.0 for all employees

---

*James F. Lincoln, *A New Approach to Industrial Economics* (Irwin-Adair Company, 1961).
[1]This is the one weakness in this plan. The performance factor should be based upon customer service, not quantity.

of several performance reviews during the year. The performance factor can range from 80 percent to 120 percent. The bonus factor has historically been around 1.0. The average of everybody's performance factor must be 100 percent. Lincoln also strongly encourages employees to purchase stock.

Summarizing the shifts in performance measurements:

	Traditional	*Enlightened*
C H A N G E D	Direct labor Efficiency Utilization Productivity	Total head count productivity Output ÷ total head count (direct, indirect, administrative personnel)
	Machine utilization	Return on net assets
	Inventory turnover or months-on-hand	Days of inventory
	Cost variances	Product cost, especially relative to competitors' costs
	Individual incentives	
	Performance to schedule	Customer service
	Promotion based on seniority	Promotion based upon increased knowledge and capability
N E W		Ideas generated Ideas implemented Lead time by product/product family Setup reduction Number of customer complaints Response time to customer feedback Machine availability Cost of quality

FINANCIAL CONTROL

As lead times become shorter and lot sizes smaller, financial controls can become much simpler. In fact, they must become simpler or the increased transaction volume, due to processing more lots more frequently, will bring even traditional computerized systems to their knees. To ensure lot control, accountants like to have numerous points at which inventory is counted. Traditional manufacturers bring parts into the stockroom and then issue them to the fabricated, rising assemblies. They count the parts when they enter and leave the stockroom. The new enlightened principles result in material flowing directly from one operation to the next, completely bypassing stores. When material flows directly, doing everything that was done before will be both impossible and unnecessary. In repetitive environments, job cost systems can be replaced by simpler process cost systems. Wall-to-wall physical inventories, which used to take several days and were extremely disruptive and costly, will take only several hours or less in enlightened environments. With little inventory, physical and financial control becomes much easier.

Overhead allocation changes drastically in enlightened environments. Personnel who run manufacturing cells become essentially fixed costs, so overhead can no longer be calculated based upon direct labor. Instead, it is based on product families' manufactured lead time or machine utilization time. Ideally, all support functions can be cellularized (assigned exclusively to a given cell). When this can be done, overhead allocation is simplest; it needs only to be accumulated and divided by production volume.

Increased purchasing transactions, because suppliers are shipping smaller lots more frequently, change traditional vendor ordering and invoicing. "Blanket" purchase orders are installed for a year. Instead of issuing purchase orders for various deliveries, one order is written for the entire year. The supplier ships to a schedule supplied periodically. With traditional practices, invoices are matched with purchase orders and receiving documents, which are generated when material is received from the vendor. When invoices, purchase orders, and receivers match, the supplier is paid. By comparison, when vendor material is used in production immediately upon receipt, and the in-house lead time is very short (i.e., several days or less), more simplification can result. The ideal is to pay the supplier based upon product shipped, eliminating the need for the supplier to invoice, receivers to be generated, and all of the matching to take place. For example, if the buyer made 1,000 products in one month, each using 10 parts from the vendor, then at the end of the month the buyer would send the vendor a check for 1,000 times 10 parts.

To do this there must be:

- One supplier for a given part.
- Extremely short manufactured lead times.
- Very few rejects (i.e., less than 10 per million).
- An unusually high degree of trust.

Understandably, satisfying these requirements will take a long time and may never be completely achieved. However, it is important to maintain a vision of what ultimately could be done and to continually press on to attain that vision. Although still on a very limited scale, more and more of these changes are being implemented in the United States today.

CHAPTER FIVE

Strategy for Manufacturing Companies

The philosophy of manufacturing excellence is of enormous strategic benefit. Its implementation can result in the following achievements:

- Much shorter new product and engineering change introduction.
- More flexibility to schedule changes.
- Significantly lower product cost.
- Lower break-even volumes.
- Significantly more organizational "stretch."
    Greater degree of profitability during all phases of the economic cycle.
    Reduced employment fluctuation.
- Better integration with customers and suppliers.
- More funds for new product and process development.
- Facilitation of "buying low and selling high."

Gaining these benefits requires a significant shift in thinking.

Traditional Thinking

Most companies have a documented business strategy, but very few have even thought about a "manufacturing strategy" per se, let alone documented it. In many cases the strategy is a single statement, such as "be the low cost producer." Representatives from the manufacturing or production function have usually been excluded from the strategic planning process. The possibility that manufacturing could provide a sustainable competitive advantage has rarely been entertained. The business strategy has been developed by strategic planning, product planners, marketers, financial managers, and general managers—many of whom have little, if any, manufacturing background or understanding. The unwritten manufacturing strategy has in essence been the following:

- *Respond to schedule changes* requested by marketing and to product changes requested by marketing and engineering. Manufacturing's charge has been "be flexible," which really means respond to just about any and all requests for schedule and product changes. Few persons understand that trying to respond to everything actually reduces flexibility overall.
- *Reduce costs by improving direct labor productivity.* Even though direct labor accounts for only 5 to 10 percent of total costs of most manufacturers, the focus of reducing costs is still on direct labor. Reduction of material (50 to 60 percent of total costs) and overhead (30 to 45 percent) is usually given secondary priority.
- *Keep inventory as low as possible.* This is frequently a secondary priority compared to direct labor productivity. The production manager is often solely accountable for all inventories (raw material, purchased parts, work in process, and finished goods) even though he or she has little control over the major driver of inventory, the master schedule. Although the master scheduler usually reports to production through the materials organization, the function in essence only maintains the master schedule in the computer. Marketing really determines what the schedule is in most plants but is rarely accountable for inventory, not even finished goods.

  Since it is not accountable, marketing tends to schedule on the high side to ensure product availability. By offering special discounts and so forth, marketing frequently creates severe fluctuations, which are virtually impossible for manufacturing to satisfy. Also, engineering generates inventory by initiating an excessive number of engineering changes. Half of these changes could be eliminated if engineering understood the production process better. Designing new products to use different parts when most products could use the same part generates inventory unnecessarily.

- *Vertically integrate* to generate the most direct labor hours. This results in greater overhead absorption cost accounting systems, which is what most manufacturers use. Lately, however, the trend is for more companies to farm out production than to vertically integrate, thus creating the "orphan plant" syndrome—less production with the same overhead drives up per unit cost, so more parts look unfavorable in a make/buy analysis.
- *React to the demand pattern;* that is, lay people off when business demand is low, and add people when demand is strong. This is especially true with direct labor personnel. The real strategy with most companies is to maintain very "tight" direct labor and "loose" indirect labor. As a result, direct labor is frequently poorly trained and has little opportunity to work on improvement.
- *Utilize a material requirements planning system to plan and control production.*
- *Purchase material with the lowest initial cost* by having at least three suppliers and playing them off against one another. Aggressively pursue foreign sourcing to obtain the lowest cost.

"Enlightened" Thinking but Devoid of JIT Understanding

In the 1970s, Wickham Skinner, professor at the Harvard Business School, emphasized the importance of retaining focus in manufacturing. Otherwise, an operation would not do anything very well. This was an important contribution because most U.S. factories had become an extremely complex collage of manufacturing processes and products. This occurred as processes and products were added piecemeal to what already existed, without considering the best organization and layout for all products. Skinner's point was that significant economies could be realized by focusing entire factories on specified products or product families. The relatively few companies that focused their factories did reap the large benefits Skinner predicted.

Skinner also asserted that an operation must focus on one of the following four strategies because it could not do more than one at a time very well:

- Be the low-cost producer.
- Offer the best customer service.

- Be the most flexible (in terms of schedule and/or product changes and new product releases).
- Have the best quality (of conformance)[1] for a given design.

If a company tried to focus on more than one of these four areas, according to Skinner, it would be at a distinct disadvantage to a competitor that wanted to be best in only one area.

Using this concept of focus, Professor Skinner proposed the following methodology for developing a manufacturing strategy:

1. Start with the business strategy.
2. Based upon the business strategy, determine which of the four areas is appropriate.
3. Evaluate the current operations capability relative to the area selected.
4. Analyze competitors relative to this focus area.
5. Based upon this competitor analysis, decide on the focus area.
6. Determine the "gap" between current capability versus desired capability. This analysis may show the gap is so big it is not practical to close it in the required time frame. In this case, the business strategy would be changed until a compatible manufacturing focus is achieved.
7. Develop appropriate policies to support the manufacturing strategy and to close the gap.

The important new concept in this methodology was that manufacturing must be an integral part of the strategic planning process and manufacturing's role might contribute the significant competitive

---

[1]Quality of conformance is the degree to which the product conforms to the specifications of its intended design. By comparison, quality of design relates to different product designs and their ability to satisfy different customer needs. An example illustrates the difference. The quality of design for a Mercedes-Benz is clearly different than the quality of design for a Honda Civic. Most people would say the quality (of design) of a Mercedes is *higher* than a Honda Civic. However, the quality of conformance of a Civic may be higher than the quality of conformance of a Mercedes. Applying the manufacturing excellence philosophy allows a simultaneous advantage in productivity, flexibility, service, and quality of conformance, not quality of design. The philosophy obviously will not allow Honda to manufacture a car with the design specifications of Mercedes more cheaply than it can manufacture a car with the specifications of a Civic.

edge. For example, relative to the competition, the company may be more responsive and able to deliver products in a shorter lead time. This factor would sustain and enlarge market share.

Enlightened manufacturing strategy. Based upon the world-class philosophy, goods factories can simultaneously accomplish low costs, high quality, minimum investment, short cycle times, high flexibility (relative to product and schedule changes), and rapid introduction. This has been proved by the great Japanese companies and the few U.S. companies that have implemented the world-class philosophy. The following major OEMs have demonstrated superiority in every dimension relative to competitors that have not embraced the spirit of manufacturing excellence: Aisan Warner, Canon, Hitachi,[2] Honda, Mitsubishi, Nikon, Toshiba,[2] Toyo Kogyo, Toyota. Many suppliers of these OEMs have also demonstrated this superiority.

To some small degree, Skinner's statement about trading off dimensions is valid. For example, by sacrificing cost, Toyota probably could increase product flexibility. However, what really matters is how the company stacks up relative to the competition. In this regard, Toyota is best in all dimensions, as are the other companies listed above. Given that simultaneous advantage is possible, is there a new methodology to replace the old one? Clearly, there is no need for the old methodology that determined an appropriate manufacturing focus area. The manufacturing strategy becomes *implement the philosophy of manufacturing excellence!*

This philosophy is applicable to continuous-process and discrete manufacturers as well as all types of operations:

- Make-to-stock.
- Assemble-to-order.
- Make-to-order.
- Engineer-to-order.

It is also applicable to all functions within a manufacturing company—engineering, marketing, finance, information systems, as well as manufacturing. One manufacturer of large valves gained a competitive advantage by applying the philosophy to "premanufacturing"

---

[2]In some of their businesses.

operations (i.e., bids and proposals, design engineering, order acceptance, marketing, production planning) to reduce lead time from 34 weeks to 8; manufacturing was left alone entirely.

The philosophy is even relevant for service industries—banking, insurance, health care, and so forth. One hospital reduced operating room capacity needs and costs by reducing changeover (i.e., setup) time by applying setup reduction concepts. Aetna's auto insurance division reduced the average lead time to process auto accident claims from 40 to 3 days.

The manufacturing strategic planning process now becomes:

- Envision what the operation can evolve to by implementing the philosophy of manufacturing excellence.
- Develop a strategy to attain the vision.
- Develop a plan to execute the strategy.

Implications of this enlightened process for most manufacturers are:

- Little, if any, additional plants, even in growth markets. Severe plant consolidation in low-growth markets because of considerably better asset utilization.
- A strategy of "simplify, then automate, then integrate," relative to automation and computer integration of the "islands" of automation. Simplification is accomplished by diligent application of the manufacturing excellence philosophy. This has clearly been the strategy of world-class Japanese manufacturers.
- Simplified computerized manufacturing control and cost accounting systems.

Conclusions

- Regardless of what business you are in, the manufacturing excellence philosophy is appropriate. It should become the manufacturing strategy.
- Even if strategic planning concludes that manufacturing is not going to offer a distinctive competitive advantage, implement the world-class philosophy.
- Simplify by utilizing the philosophy, then automate, then integrate.

The following are implications of the philosophy as compared to the traditional manufacturing strategy discussed earlier:

Response to schedule changes. The operation will be more responsive as a result of lead times being reduced 70 to 90 percent. However, accomplishing this in most operations requires what may seem to be a paradox; some schedule stability for 2 to 4 weeks is required to reduce overall average lead times. Following is a "before" and "after" scenario of a medical instrumentation manufacturer:

	Before	*After*
Schedule change policy	Change the schedule to agree with marketing demand in the immediate time frame if need be	Month 0–1 do not change
Average lead time for a significant schedule increase	8 months	1–2 ± 10% 2–3 ± 20 3–4 ± 40

The "after" scenario took several years of significant change at the OEM and key suppliers.

Skeptics of the success of great Japanese companies say the Japanese success is due to their ability to freeze the schedule for many months and this would be impossible in the United States; therefore, the philosophy is unworkable because U.S. manufacturers must be more flexible. Actually, the Japanese companies are much more flexible overall. They freeze the schedule, but usually for only one to two weeks to avoid the waste of confusion. However, no company, not even the great Japanese companies, can afford to ignore a customer in trouble. They will respond within a short time, but do so in a much more disciplined way than in the United States. Freezing the schedule allows the schedule to be leveled, which in turn allows production of all parts to be synchronized with the level schedule. Lead times can be reduced at a rate that is directly proportional to the degree to which all schedules are synchronized.

Reducing costs. Cost reduction efforts focus on material and especially overhead because it is controllable. By delegating more authority to direct labor to streamline the flow in the shop, the overhead required to support the shop can be reduced significantly. Direct labor may actually need to increase if it is to be involved sufficiently.

Inventory performance. Inventory is seen as a reflection of how effective the entire operation is, including marketing, engineering, and manufacturing. All of these functions share in the rewards of good or bad inventory performance. The focus is on the improving production processes. Inventory reduction, per se, is not a focus, but is viewed as a natural by-product of improving the processes.

Engineering strives to design new products using common building blocks or modules to limit the number of parts and inventory but still have product flexibility.

Vertical integration. Vertical integration is no longer pursued because it adds direct labor hours so more overhead can be absorbed. This is due to several reasons, including:

1. Return on assets (ROA) is the key objective. If parts can be purchased at the least total cost (includes service, quality, and lead time), then, everything else being equal, it does not make sense to have those assets in-house. Furthermore, if ROA is the overriding criterion, generating unrecorded inventory will offset the benefit to profits.

2. Since overhead is reduced significantly, it is not as much of an issue. Break-even volumes can be reduced to 20 to 30 percent of capacity.

Vertical integration will tend to be considered more for other reasons. If an operation is a JIT producer, it can make parts as effectively as anyone else. Therefore, why shouldn't the parts be manufactured in-house? Also, applying the manufacturing excellence philosophy reduces head count requirements drastically, potentially much faster than new business can be obtained to offset the reduction. Making instead of buying helps provide employment security.

React to the demand pattern. The philosophy provides greater organizational "stretch"; that is, a smaller number of people can produce profitably through a wider range of demand. Instead of maintaining "tight" direct labor head count and "loose" indirect, enlightened companies run more with "loose" direct labor and "tight" overhead.

Simplified MRP. MRP systems are not obsoleted by the philosophy, but become much simpler. Enlightened companies focus on simplifying the processes, realizing that systems needs will become simpler also. This approach contrasts with traditional thinking, which assumes the process is what it is and tries to use a system to optimize around problems.

Supplier consolidation. The number of suppliers is consolidated significantly to:

1. Become a more significant customer to suppliers.
2. Be able to relate to suppliers.
3. Reduce the variation of purchased material arising from having many vendors.
4. Reduce administrative costs from dealing with many suppliers.

Enlightened thinking works toward establishing more trusting, longer-term relationships with suppliers that are as close to the manufacturers as possible. As smaller quantities are shipped more frequently, transportation costs become more critical.

Closeness facilitates communication and shortens the logistical pipeline. As the logistical pipeline shortens, total costs are reduced proportionately, especially inventory costs. The traditional "knee-jerk" response to buy anywhere in the world when the correct initial cost can be found is being replaced by a greater sensitivity to total costs, including quality, inventory, service/response, and logistics. Every attempt is made to develop suppliers as close to the OEM as possible. Consequently, the stampede to foreign sources is abating.

CHAPTER SIX

Strategies for Implementing Technology in Manufacturing

Introduction

Many Americans claim we cannot "out-Japanese" the Japanese; that is, the United States cannot compete with Japan by being better disciplined, managed, or organized. Instead, they contend the United States must utilize technology to leapfrog the Japanese. By technology, they mean more automation and computer integration—specifically, computer integrated manufacturing (CIM), which is largely comprised of flexible manufacturing systems (FMS). The rush to CIM is strongly echoed by machine tool manufacturers and computer hardware and software companies.

Magazine and journal articles about highly automated plants, such as MAZAK's in Florence, Kentucky, imply that manufacturers must automate to the degree that MAZAK has in order to survive. This implication may mislead and damage many U.S. manufacturers for several reasons:

1. Many do not have the millions of dollars that MAZAK's manufacturing systems required. Some manufacturers may throw in the towel when they hear this message. Fortunately, most benefits that MAZAK reaped by automating can be achieved by investing only a fraction of the capital spent by MAZAK. This plant utilizes FMS, comprised of computer numerically controlled machine tools and automated material handling devices. All machining operations and material handling are controlled by a central host computer. The FMS is not appropriate for many U.S. manufacturers. Most benefits can be attained at a fraction of the cost of an FMS, which usually ranges from $5 million to $20 million. The MAZAK plant is a collection of FMS, which raises the cost even higher. Most benefits are attainable by simply employing the world-class philosophy and techniques.

2. Even if a manufacturer can afford the systems, it may not be ready to implement them. In order to successfully implement these systems, the environment must first attain a high degree of discipline. Some companies justify the expensive automation by benefits attainable at a fraction of the cost by simply gaining control of their shop floor without automation. These companies may not be able to justify this expensive automation when evaluated after all the benefits of control are attained.

In the past 15 years, U.S. industry has spent approximately $20 billion on computerized manufacturing control systems, called MRP. Few work well because they are in environments that are not well enough controlled. These MRP systems are less complex than what exists at MAZAK.

Most people believe the major hurdle in implementing CIM is integration of the myriad of different hardware and software used by the various "islands of automation." But this problem is small compared to the task of automating production processes with extreme degrees of variability, the predominant type today.

MAZAK's degree of automation is not required by *all* manufacturers to be competitive. What MAZAK has is appropriate for *some* businesses, but only if they are adequately prepared. Most U.S. industry needs to first implement the basics and then consider the degree of automation MAZAK has. By employing the manufacturing excellence philosophy, companies not only prepare the way for future automation, but they also generate sufficient cash from improvement to afford it.

This chapter will present: (1) different strategic approaches to automation, (2) principles for selecting the right degree of technology, and (3) justification of investments.

STRATEGIC APPROACHES TO AUTOMATION

The spectrum of strategic approaches to CIM is analogous to that of the strategic approaches to football. At one end of the spectrum is "grinding it out." At the other end is "going for the long bomb."

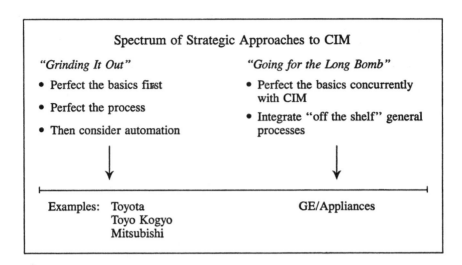

"Grinding it out" entails mastering the elements of manufacturing excellence (e.g., managing random process variation so it is significantly less than the specification requirements, reducing setups to less than the specification requirements, reducing setups to less than 10 minutes, using preventive maintenance to ensure machine uptime is greater than 99 percent, and so on) to get the most out of existing equipment. It means simplifying to the highest possible degree first. Next, further automation is considered and, when appropriate, automation is integrated. Improving the process by internal capability is emphasized. In-house personnel adapt equipment to the needs of the specific process. As a result, the adapted equipment cannot be purchased externally. It becomes a proprietary process and, as such, a competitive advantage. The credo of those who grind it out is "simplify, automate, then integrate." This approach requires much time and that many small steps be taken every day.

"Going for the long bomb" means simplifying, automating, and integrating simultaneously. Instead of custom designing processes, the firm purchases off-the-shelf equipment. The focus is on integrating general purpose equipment and using the FMS. Frequently, the rationale behind implementing complex FMS's is: "Our strategy is to utilize state-of-the-art technology. FMS's are the current state of the art and the sooner installed the sooner we'll have a competitive advantage. Also, since FMS's are integrated in themselves and computer controlled, a significant part of the ultimate vision (i.e., CIM) will be in place."

FIGURE 6-1

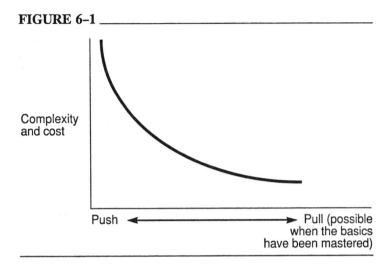

Comparing the Strategic Approaches

Complexity and cost. Complexity and cost go hand in hand. The more complex, the greater the cost. When the basics of manufacturing excellence are in place, the environment is much simpler. Quality problems are rare. Processes are in statistical control and, therefore, are predictable. Process variation is considerably less than specifications. Setups are 10 minutes or less. Machine uptime exceeds 99 percent. There is little work-in-process inventory. CIM is considerably harder and, consequently, much more costly when the environment has not been simplified to the furthest degree. Figure 6-1 shows that complexity and cost decrease as an environment evolves from "push" to "pull." This is possible in a repetitive environment when the basics are in place and does not require a computerized shop floor control system.

Ease of implementation. "Grinding it out" is simpler to implement because it is done in small incremental steps over a long period of time, whereas the "long bomb" tries to do everything—the basics, automation, and integration—in a short time.

Break-even volume. "Grinding it out" does everything with minimal capital first, then considers automation. The "long bomb" requires a large capital investment and, therefore, has a high break-even volume.

Risk. Because the break-even point is higher and implementation occurs over a shorter period, the "long bomb" approach entails more risk. The chances of successful implementation are less, and, when achieved, the projected volume may not materialize. When companies implement basics and sophisticated automation at the same time, the relatively mundane basics are usually outprioritized by the more glamorous automation. Some manufacturers try to automate without any consideration of the basics. The results are disastrous.

General Motors' new $600 million assembly plant in Hamtramck, Michigan, was designed to be a showcase for automated high technology. The plant has 260 robots for assembly and painting, 50 automated guided vehicles to ferry parts, and a battery of cameras and computers that use laser beams to inspect and control the manufacturing process. Unfortunately, Hamtramck is turning out only 30 cars an hour, far less than the 60 an hour it was designed to build. This pilot "factory of the future" points out the inadequacy of automating before replacing the underlying manufacturing system with a better system. Its equipment is so sophisticated that either it breaks down on its own or workers cannot operate it, even after months of training. Automation, in itself, has not massively improved productivity.

Automation, like strong medicine, must be taken in carefully measured doses. Japanese firms use automation but in smaller doses. A new Mazda Motors plant, also in Michigan, costs 25 percent less than Hamtramck because it lacks most of GM's sophisticated automation. Yet the plant is producing 240,000 cars a year with 3,500 workers, while Hamtramck has 5,000 workers and aims at only 220,000 cars a year when everything works right. A Honda plant in Marysville, Ohio, has only two automated guided vehicles. The lesson is that with just the right amount of technology, lower investment and higher productivity result. Manufacturers can make bigger gains by implementing the basics of manufacturing excellence (JIT, TQC, and EI). Automation will cut costs and increase quality in the long run and provide such product flexibility that industry will look like a manufacturing boutique. This will be economically feasible only if the spirit of manufacturing excellence takes hold.

The "grind it out" approach achieves the benefits of automation while lowering the break-even volume.

Relative benefits. If CIM is appropriate for the business and is successfully installed, benefits from the "long bomb" could be greater and achieved earlier. If a company successfully invests in CIM, it could gain an advantage over other competitors. GE Appliances did this with its Louisville factory, which cost between $70 million and $150 million, depending on the estimate. However, the benefits could have been realized with much less investment by doing the basics better.

Appropriation by the competition. The "long bomb" approach requires general "off-the-shelf" systems. "Grinding it out" focuses on using in-house capabilities to design and implement customized processes, which cannot be replicated by the competition and are, therefore, a competitive advantage. In the early 1980s, the Toyota small engine plant at Kamigo, using 1960 vintage American machine tools, was compared with similar Chrysler and Ford engine plants using state-of-the-art machine tools. In terms of engines per total head count (direct, indirect, and administrative) the Toyota plant had 4.5 times the productivity. The Toyota plant had custom designed and perfected its processes. No one could go out and buy it.

A comparison of the two strategic approaches follows:

Comparison of Strategic Approaches

Criteria	"Grinding It Out"	"Going for the Long Bomb"
• Complexity and cost	Less	More
• Ease of implementation	Easier	Harder
• Break-even volume	Lower	Higher
• Risk	Less	More
• Relative benefits	Possibly less initially	Possibly more sooner
• Appropriation by the competition	Difficult	Easier

Both strategies can work, but "grinding it out" is desirable because it ensures that inefficiencies are not automated and benefits from doing the basics well are not used to justify high-cost automation.

American industries have been chastised severely, and rightly so in some cases, that lack of investment has caused their demise. The steel industry is an example. Consequently, industries are prodded to invest in so-called state-of-the-art equipment. Unfortunately, some companies with the money to spend have done so before preparing adequately, causing them to fail or spend money unnecessarily. There is no way around doing the basics well. They cannot be shortchanged if automation and integration are to succeed.

To review the basics, envision the plant comprised of many, many small cells. Physically create the cells as they would be arranged if the entire plant was a collection of small cells. Perfect the cell by balancing the load between operations comprising the cell. Attain single-digit setup time (i.e., less than 10 minutes) on equipment on which more than one part is run. Ensure that the processes in the cell are within statistical process control (that is, variation is random and not due to assignable causes), and that machine availability when required is more than 99 percent where processes are consistently well within specifications. When cells are this efficient, consider computer integrated manufacturing.

PRINCIPLES FOR SELECTING THE RIGHT TECHNOLOGY

1. *"Off-the-shelf" technology will not give a sustainable competitive advantage because everyone can buy it.* It is what goes around the machine that counts, not the machine itself. Things "around the machine" are:

- Integration with other operations and manufacturing cells.
- Tool handling devices and procedures.
- Material flow.
- Process control.
- Fail-safe mechanisms and processes.

As mentioned previously, a Toyota engine plant using 1960 vintage machine tools has four times the productivity of automated Ford plants. Through outstanding industrial and methods engineering, Toyota has created a superior system, even though its machinery is not state of the art. U.S. engineers focus on optimizing the machine run time and ignore the production process between the machines. Toyota, on the other hand, looks at the whole process and optimizes everything around the whole process. Frequently, older equipment runs fast enough.

2. *General "off-the-shelf" processes are more expensive than custom-designed processes.* A recent University of Vienna study showed that on average, off-the-shelf processes are at least 30 percent more expensive than custom-designed processes. Custom designed can mean designed by in-house personnel, not necessarily custom designed by an external source. The situation is analogous to computer software. Something general in application must accommodate many different situations. Consequently, it is not as efficient, in terms of total run time, as a program designed specifically for a given process.

3. *Automate material handling last for several reasons.* Automation of material handling reduces the flexibility to go from one family of parts to another. Also, the incremental benefit of automating material handling may not be justified by the incremental cost. Except in situations where the parts are so big that a mechanical conveyance is required, Japanese plants tend not to automate their flexible machining systems because they have found the cost is not justified by the incremental benefit.

4. *When planning for a new product, replicate small cells as demand increases, instead of having one larger cell or flexible machining system for the "most likely" demand.* This provides low break-even volume at all stages of growth. The U.S. tendency has been to pick a most likely volume, or new product, and to design the optimum system around that volume. Today, many people think that the optimum system is an FMS. Actually, a flexible machining system is very expensive and results in a relatively high break-even volume. If the volume of the new product is significantly less than the forecasted volume, the company will be stuck with a high fixed cost and a high break-even volume. Therefore, it makes sense as the new product is being launched to start with a very small cell comprised of less complex and less costly equipment and as volume increases to replicate these. For example, at peak demand, 10 of these small cells may be replicated. It is possible that at peak demand these 10 cells will not be as productive as FMS's designed for this volume. However, if this demand is not realized, greater flexibility will be worth the slight disadvantage in productivity at peak demand.

5. *Use the smallest possible building blocks in the cell to provide the greatest flexibility to accommodate product changes.* Small building blocks allow cells to be rearranged easily and with great variety. A machining center is a computer numerically controlled machine capable of performing most of the basic machining functions, such as drilling, tapping, milling, and so on. Consequently, it has been given the name "machining center" because it does most of the basic machining functions. It is a very flexible machine. However, using these machining centers as the basic building blocks in factories increases fixed cost too much and raises break-even volume. Again, 10 small cells are usually better than one large cell because they afford greater flexibility during the growth of demand. Having a factory full of machining centers provides great flexibility but at a high cost, especially for repetitive processes. If the particular environment's output is one of a kind, requiring much engineering change, then having a machining center is advantageous to respond to demand. However, if the demand is repetitive, much smaller cells, comprised of much less complex equipment, make sense. Manual machines may offer the lowest cost alternative. With complex parts and/or frequent engineering changes, machining centers are the obvious choice.

Some people want one large flexible machining system that supplies many product lines at the same time. But this destroys any possible product focus. Focusing equipment by product family affords the simplest logistics/material flow and so should be retained if possible. Large FMS's that supply many product lines will destroy this focus. Product focus is a key feature of the Toyota plant that has four times the productivity of its U.S. counterparts.

6. *Return on assets should be the key measure in justifying equipment, not machine utilization.* Initially, this sounds contradictory because high machine utilization has been associated with high return on assets. Return on assets is a function of asset utilization times asset value. By using lower cost equipment, a lower asset utilization can be offset and still provide contingency capacity for stretch. Planning to utilize machining centers for three shifts allows no contingency. Many people invest in very expensive machining centers. To maximize machine utilization of this equipment, it is run around the clock for three shifts.

This scheme includes no contingency planning for problems and no time for preventive maintenance. Over time, the equipment deteriorates and breaks down. Frequently, if demand is low, unneeded inventory will be generated. The real key in evaluating assets and equipment is the return on total assets: inventory, as well as machines, plant, and so forth. It is asset utilization times the asset value.

Consider dedicating fully depreciated equipment to machining specific parts or utilizing this written-off equipment in cells. With the entire process in mind, no operation should go faster than the slowest link in the total process chain, because if one operation goes faster than another, inventory appears, the worst form of waste because it covers up problems. Consequently, that old equipment may be fast enough. It may be as fast as the slowest link in the total chain. Generally, fast equipment is difficult to use when you design with the whole process in mind because the faster equipment generates inventory.

7. *When purchasing equipment, the deciding factor should be the equipment's effect on overall reliability of the system.* Managers have traditionally purchased equipment with the lowest cost and have not considered its impact on the system's reliability. If that machine goes down, the cost of the whole system going down could more than offset any savings on the low-cost machine. Campbell Soup Company now purchases equipment not with the lowest cost, but equipment that offers the greatest reliability. To Campbell, reliability not only means running when it is desired, but also being able to repeat its operation within the desired tolerance band.

8. *Select equipment that allows product focus.* As discussed previously, product focus is ideal because it minimizes logistics, which results in the fastest throughput. Throughput time should be a driving factor because the faster the throughput, the greater the productivity and the better the quality. Furthermore, arranging things in a product focus will result in equipment and processes having close physical proximity, which allows for control by visibility. When this is possible, a complex computer status and control system is not required and the resulting overhead saved.

9. *People allow the greatest degree of flexibility.*

It is important to point out that some factory components should be avoided as much as possible, such as:

- *Automated storage and retrieval systems (ASRS) to store inventory.* These automated storage systems cost well over $1 million. Rather than spend money automating the control of stored material, energy should be spent figuring out how to eliminate inventory. Automated storage and retrieval systems in the context of the world-class philosophy become dinosaurs and monuments to inefficiency.
- *Automated guided vehicles.* These are disadvantageous for several reasons. First, they are limited by a fixed route. Second,

they imply that things will be moved in batches. The goal of JIT is to move things one piece at a time, not in batches. An automated guided vehicle will make moving in batches more efficient, but it becomes a very unproductive means of material conveyance when the goal is to hand off parts one at a time between machines that are next to each other.

- *Forklift trucks inside the factory.* Forklifts are very expensive in themselves, expensive to maintain, and require operators. Material handling does not add value and, therefore, is waste. Forklifts, like automated guided vehicles, imply moving things in batches and, therefore, should be designed out of the system. A goal of every operation should be to eliminate forklifts inside the factory. Operations should be close enough and lot sizes small enough so parts can be moved one at a time, even in very high-volume, repetitive environments. Use conveyors to move material from one operation to another. However, using conveyors for storing material is wasteful. Conveyors should be used only as a means of material conveyance.

- *Computerized shop floor control systems in repetitive environments.* In a repetitive JIT environment, computerized shop floor control makes no sense. The purpose of computerized shop floor control is to release orders to the shop and to provide status of the orders when they are in the shop. In a repetitive JIT environment, parts move so fast it is unnecessary to keep track of them. Control (i.e., authorization to move and produce parts) is made by sight. A computerized system to do this is redundant in such an environment. In nonrepetitive shops, it is required. The problem is defining what is and is not repetitive. Many traditional manufacturers think they are nonrepetitive, when in fact much is repetitive after the JIT philosophy has been applied.

- *Automated material handling between machining centers.* As mentioned earlier, automated material handling reduces flexibility; that is, it tends to be designed for a certain size family of parts. If the family changes, it is very costly to change the equipment. Also, the additional benefit derived from automated material handling equipment usually does not justify the cost. There are, of course, examples where automating material handling is very practical.

JUSTIFICATION OF TECHNOLOGY

Traditionally, new technology has been justified by how much direct labor it saves. Although technology continues to reduce direct labor requirements, this cost factor has become a rather small component for most manufacturers (ranging from 5 to 10 percent of total cost for the majority). Technology also affects the following cost factors:

- *Quality,* through integrating inspection at the source, using a fail-safe process, and reducing rework and scrap.
- *Setup costs.*
- *Reduced lead time,* which reduces work-in-process inventory and, in itself, can become a competitive advantage.
- *Overhead,* through reducing material handling.
- *Work-in-process and finished-goods inventories,* which are a function of reduced throughput time from greater flexibility and shorter setups.

The challenge is to appropriately quantify the affected factors. It is also important to determine to what degree the basics are in place and what the incremental advantage will be in further implementing the basics.

Consider a power transmission manufacturer that wanted to replace two manual turret lathes, two manual drill presses, one manual milling machine, and an old computer numerically controlled horizontal milling machine with one new machining center. The new machine center cost $650,000, including new tooling, installation, and training. The initial justification was as follows:

		Annual Savings
Setup	Reduced by 90%. As a result, lot sizes could be reduced from monthly to weekly, which would increase the number of setups by 400%. Total annual setup time would be 400% × 10% = 40% of what it was previously.	$ 80,000
Quality	Rework and scrap reduced from 10% and 5% to 2.5% and 1.25%, directly proportional to the percentage of lot size reduction (i.e., 75%). (They reasoned correctly that rework and scrap would be reduced proportionally to the lot size reduction because with less in process, you make less scrap and rework before you find out there is a problem.)	250,000

		Annual Savings
Work-in-process	Reduced by 70% due to shortened lead time and reduced lot size from one month to one week $700,000 × 70% × 24%.	$125,000
Direct labor	Reduced by 3 persons.	120,000
Maintenance	The maintenance supervisor estimated a savings due to having one new piece of equipment, instead of six older ones (based upon the previous two years of maintenance for the new equipment). Also, it would be easier to deal with one equipment supplier instead of three.	100,000
Overtime (O/T)	Reduced direct O/T by $100,000 and indirect (for expediting) $40,000.	140,000
Reduced purchased parts premium cost and freight premium	To respond to schedule change requests.	120,000
	Total	$935,000

By comparing the projected annual savings of $935,000 with the cost of $650,000, this appeared to be an outstanding investment—approximately nine months' payback and a very high ROI, even assuming a short economic life and a 20 percent cost of capital. In addition, the manufacturer believed it satisfied the company's manufacturing strategy, which was "to utilize state-of-the-art technology." Furthermore, one CNC machine would be easier to integrate into a CIM network, the manufacturer's goal. Considering everything, the investment looked like a "no brainer."

An evaluation of this investment in light of the JIT philosophy follows:

	Evaluation	*Annual Savings Compared to the Operation with the JIT Basics in Place*
Setup	Little had been done to reduce setup with existing equipment. Based upon experience with this machinery, a 90% reduction was possible with very little capital investment ($5,000).	$ 0

Evaluation		*Annual Savings Compared to the Operation with the JIT Basics in Place*
Quality	The quality savings identified would follow from setup reduction and do not require additional investment. There was no systematic evaluation of process capabilities compared to specification requirements. If there were, significant quality improvement would be possible without capital expenditures beyond what was attributed to the new machine.	$ 0
Work-in-process	The evaluation of the impact of WIP did not consider that having one machine supply three product lines complicated the material flow more than the current arrangement. As a result, some WIP reduction was possible, but not as much as estimated. By utilizing another piece of equipment that was going to be disposed of, by purchasing another small milling machine for $30,000 and by rearranging existing equipment (including the machines the machining center was to replace), the WIP could be reduced by at least 90%. Compared to an operation with the basics in place, there would be $100,000 more inventory, the carrying cost of which was $25,000.	25,000
	It is important to point out that the 25% used to calculate annual inventory carrying cost is grossly understated. Given that inventory masks many problems, 48% is a more appropriate number.	
	Further, machining center breakdown would cause an enormous logjam because it was now a very large bottleneck.	
Direct labor	Reduction of three persons will follow from doing the basics well due to reduced material handling time and "hunt and search" time for parts, tools, gauges.	0
Maintenance	Preventive maintenance, another basic, was not done at all. The equipment was essentially run until it broke. None of the equipment appeared to be in need of such drastic repair that it would be better to junk it.	0
Overtime and purchased parts and freight premium	Possible from doing the basics well.	$25,000

By making a $30,000 capital expenditure, plus $5,000 for costs related to setup reduction on existing equipment, annual savings would be $25,000 more than if a $935,000 expenditure was made. Stated another way, relative to what the operation could be like with the basics in place, the machining center investment resulted in a $25,000 loss!

The manager of industrial engineering, who was the champion of the machining center, initially had difficulty accepting the new analysis. However, after several setup reduction plant projects reduced setup by 85 percent without capital expenditure, he started to see the light. He finally stated that even if the analysis were true, the firm would be better off in the long run with the machining center because it kept the firm near the state of the art and better able to implement CIM. While this was true to some degree, it was not significant enough to offset the cost disadvantage.

In some cases, new technology is a significant advantage. Continuous casting is clearly superior to discrete ingot casting for the steel industry. However, when it comes to machine tools, the correct answer is frequently not so obvious. Many times fast equipment does not fit because it is too fast relative to the other operations in the total pipeline (from raw materials and purchasing parts to completed items). *The latest technology must be continually evaluated.* But it is overly simplistic to say the latest development is state of the art, and state of the art will give a company a competitive advantage. Do not acquire technology for technology's sake. If the basics are not in place, the new technology may not be necessary, and probably will not be utilized enough because of excessive quality problems and machine downtime. Integrating equipment requires that each machine be highly reliable (99.9 percent)—if one machine goes down, the whole integrated system goes down. To avoid this problem, one defense contractor actually added a redundant machining center and an extra automated guided vehicle. Needless to say, these additions raised the fixed investment enormously. The company spent approximately $10 million on this FMS unnecessarily. By arranging existing equipment into cells and reducing setup times, no investment was required!

Usually, equipment is justified individually without considering the entire flow. In order to justify equipment properly:

1. Envision the ideal operation as a collection of product-focused cells. It may be impractical to do this because asset

utilization is too low. Because of practical considerations, back off from this ideal when appropriate to arrive at a vision that is product focused to the furthest practical degree.

2. Develop a strategy to implement the vision.
3. Justify the total vision as a whole. As specific equipment is purchased, do not rejustify individually. Instead, ensure that the individual equipment is consistent with the vision and strategy. Justification based solely on replacement of worn out equipment is totally unacceptable.

Remember the different biases of the company's other functional areas and of its suppliers of goods and services.

Industrial engineers (IEs) are biased toward CIM, which most of their publications tout as the factory of or with a future. Few IEs understand the degree to which the basics must be in place first or have much experience with the basics. Instead of improving the methods, many of today's IEs would rather work on systems that optimize around bad processes. Integrating the manufacturing excellence philosophy requires a heavy investment in internal methods engineering capability.

Materials management and *cost accounting* people are biased toward computerized manufacturing control systems, which cannot improve the process; they only optimize around problems. Of today's materials and cost accounting organizations, 30 to 70 percent are not required in an organization that has evolved sufficiently with the manufacturing excellence philosophy.

Engineering is biased toward perfecting the design, frequently suboptimizing it relative to productibility. Few people consider the whole process from perception of customer need, to manufacturer, to supplier, and work to improve that whole process. The world-class philosophy requires a renaissance of methods engineering thinking by everyone from the top floor to the shop floor, including design engineers. An extremely strong methods engineering capability is required to implement technology satisfactorily.

Conclusion

In summary, American manufacturers must innovate by implementing an entirely new philosophy of manufacturing to compete with the world-class firms of Japan. Relying on automation alone will not provide the competitive advantage needed in quality and productivity.

This philosophy of manufacturing excellence, which includes just-in-time and total quality control, provides the basic concepts and techniques that will allow for effective automation.

At this time, automation most often should be taken in small doses and will improve productivity to some degree. In a few cases, automating entire operations may be possible but at significant risk and investment. The "grind it out" approach is frequently preferable because it ensures the basics become a permanent part of the firm's culture, inefficiencies are not automated, and needless automation is avoided.

Principles for selecting the right technology, automated or otherwise, have been developed and implemented successfully in American plants. As a result, many traditional concepts and practices no longer work. High reliability should replace low cost as a deciding factor in purchasing equipment. Return on total assets, not machine utilization, is key to evaluating an operation's performance. Other new approaches include emphasizing in-house development of internal processes rather than purchasing off-the-shelf processes, stressing small machine skills rather than FMS's, and using the product focus approach to minimize logistics.

A global perspective of the entire process flow should guide all decisions and purchases. Manufacturing excellence favors eliminating equipment that, from an enlightened viewpoint, reduces productivity. If a firm commits the necessary energy and time to the new philosophy, it may gain the full benefits of automation without the tremendous capital investment. After the basic philosophy is in place, the arrival of the factory of the future will be only a matter of time.

PART THREE

Implementation

CHAPTER SEVEN

Changing the Way the Company Works— A Road Map for Change

One CEO of a Fortune 500 company said, "I understand manufacturing excellence and want it. How do we engage everyone in the corporation so that they think this way? Obviously, everyone must be on board with something this profound. In many ways this is a cultural change. How do we change culture?" This chapter answers the CEO's question by describing how to implement the manufacturing excellence philosophy.

TOP MANAGEMENT'S ROLE

Cultural change amounts to changing some basic long-held beliefs and values. Because of this, cultural change is difficult and usually takes years to accomplish.

It must start with the top manager. Only if the top manager's behavior consistently reflects the new values will lasting change occur. An operator stopping production when there is a quality problem is a classic example. Traditionally, doing this would have resulted in at least a severe reprimand. Very high-level support is required for someone to do this and be applauded, not crucified. Even when the top person's behavior does reflect new values, it usually takes a lot of time for most people to really believe there has been a change and even longer for employees to change their own thinking. In large organizations, significant progress can be realized in the bowels of the operating units before the CEO even hears about what is going on. However, in order for a company to become a world-class competitor utilizing the manufacturing excellence philosophy, its top manager must proactively embrace and behave in a way consistent with the new philosophy.

To this end, the top manager must:[1]

- Expect the development of a manufacturing strategy if it doesn't already exist.
- Ensure the successful integration of the manufacturing strategy with existing engineering, marketing, and financial strategies to facilitate attainment of the business paradigm.
- Persist at breaking down functional walls that are historical barriers to business success.
- Solve the continuity problem caused by promotion of people successful in the short term, as pursuing manufacturing excellence is a long term journey and not just a high priority item in the current year's MBO program.
- Establish breakthrough expectations/goals for the organization. This is one of the all too little reported secrets of the Japanese success. Japanese CEO's establish reach-out goals for their employees that are not negotiable.
- Develop and widely communicate a commitment and plan for the excess people that will result from manufacturing excellence progress.
- Personally conduct a Total Quality Control audit to assess the change process to institutionalize the new thinking throughout the company.

Allan Cox, business columnist of the *Chicago Tribune,* has stated:

> At the risk of appearing incredibly closed minded, or just stupid, I would like to offer my view that the reason *most organizations leap to adopt the "latest advances," unwittingly of course, is to spare themselves the painful commitment to the basics.*
>
> Staying abreast of technology is essential to the growth of any enterprise, but it needs to be said that the application and usefulness of technology is almost always evolutionary, rather than revolutionary. After all, the computer has been with us more than 40 years, and we're just coming to trust it.

---

[1]Contributed by my good friend Ken Stork, Corporate Director of Materials and Purchasing for Motorola and Chairman of the Association for Manufacturing Excellence.

I also think that all the talk about "unprecedented change" is subterfuge. First of all, most of this change was foreseen years ago by perceptive executives in all industries. Second, these "seers" were not heeded by others, who, were they to respond appropriately, *would have had to commit themselves anew to fundamentals.*

In the end, we falter not for failure of nerve in the face of crisis, but for refusing accountability in the face of evolutionary change. Evolutionary change gives us an abundance of time to perfect our crafts. If only we would!

True distinction in business usually does not lie in the exotic. It lies in making the seemingly exotic mundane through drill and drudgery.

Look around your company again to reassess the stellar performers. Be suspicious of those who talk about the new era. Rather, *get behind those who have delivered year in and year out.*

Myron Tribus, director for the Center of Advanced Engineering Study at MIT, suggests:

A manager is responsible to give consistency of purpose and continuity to the organization.

The workers work *in* a system. The manager should work *on* the system to see that it produces the highest quality product at the lowest possible cost.

Accepting the fact that top management's role is not solely one of dealing with strategies or "global" matters but of improving the management system is the first step in becoming involved.

Tribus provides some good insight into how top managers might need to change their own "systems."

If you observe closely the decision process in a large corporation, you will (typically) find that the CEO is not really making the decisions he thinks he is. The alternatives were developed by the staff and the system. The choices placed before the CEO were analyzed by many people, each with specialized knowledge and interests. The alternatives are ranked by various numbers of dubious validity, bearing on what the future will bring under uncertain circumstances. *If the CEO gives either approval or disapproval, his judgment applies to the processes that produced the information, not really to the decision itself.* Only in a small company does the CEO know enough to make a decision without the aid of a staff. *If the CEO does his job properly, he will spend time making sure the organization understands its purposes and then will devote his energies to improving the ability of the system of people and machines to make these decisions.*

Top management's time must be organized to give high priority to achieving manufacturing excellence and working *on* the system to provide the mode for accomplishing this.

In turn, subordinate manager's time must also be prioritized accordingly. Recognizing and accepting this fact of time management is the first step toward commitment. It is followed by developing a more than superficial understanding of what manufacturing excellence is. You have begun already by reading this book about a never-ending, sometimes painful commitment, one that Allan Cox believed is too painful for most organizations to continue with.

One example of top management's lack of understanding recently occurred in a very successful U.S. company where several upper-middle-level line management people understand manufacturing excellence and have achieved substantial benefits from implementing its philosophy. However, top management and corporate staff functions are just beginning to learn about it.

One such champion within the company has reduced inventory and indirect costs significantly. But the substantial inventory reduction has resulted in underabsorption of overhead and is a concern of accounting. Top management, on the other hand, is worried about efficiency (direct labor is only 10 percent of the product cost), which has been improved but not as substantially as the other two areas mentioned. The top managers acknowledge there have been improvements, but these improvements are not readily reflected in the current measurement factors and they are having difficulty understanding this.

To compound this situation, the champion is requesting money for employee education instead of more capital investment. The champion knows that only a few of his employees are involved in improving the system (or process), leaving a significant untapped opportunity for more substantial benefits. Top management, while sympathetic, is having a difficult time reconciling this. To their credit, they are now taking time to understand this new set of circumstances and its applicability across the company.

This whole situation would not have occurred and the improvement process slowed, if top management would have been involved initially and understood the fundamental issues. Other companies have experienced worse results. Many middle-level managers have championed manufacturing excellence in their domain of responsibility only to be rebuffed by top managers, who did not take time to understand the significance of the changes and become involved. Instead, they took the easy way out and kept their status quo. This is the painful part of the commitment. Not only does the "system" change for the workers but also for top management "actions" too. The significance of this commitment though is probably the difference between whether the company remains competitive in today's global economy.

CHANGE STRATEGY—"GO WITH THE FLOW"

Implementing significant change is a function of *leadership capability* to provide direction, inspiration, and support, *knowledge* about what to do and how to do it, and the *enthusiasm, will, and energy* to do what is required.

Change = f (Leadership, Knowledge, Enthusiasm/Will/Energy)

"Going with the flow" means identifying those persons who have the right kind of leadership (i.e., that which promotes employee involvement) and who have the enthusiasm, will, and energy to want to do something. Knowledge can be provided.

The following leadership profile exists in most companies:

Categories	Percent of the Leadership
Movers and shakers	10–15%
Those waiting to see if this is just another program or if it really has staying power	80
Those who will resist the change forever	5–10

Let everyone know about the philosophy of manufacturing excellence. Identify those who are enthusiastic enough to do something. Provide support and resources to the movers and shakers. Use their successes to develop further commitment and to motivate others. Don't spend a lot of energy trying to get the wait-and-see group to do something. Keep the others informed about what the movers and shakers are doing. The long-range goal is to change the profile so more leadership falls into the movers and shakers category.

STAGES OF THE MANUFACTURING EXCELLENCE EVOLUTION

Many organizations seem to go through the following stages in the evolution, but it is not essential to evolve in this sequence. Some top managers become zealots without needing to see some of the concepts implemented in their company and essentially enter the evolution at Stage VI.

Getting Started

Start by teaching the company about manufacturing excellence, giving examples of how various elements of the philosophy have been implemented in other companies. Then expect various plants/operating groups to do something in three to six months with a maximum payback of six months; this will preclude alternatives requiring large capital expenditures.

The philosophy is best learned by doing. Most people need to be involved with it for three to six months in order to appreciate its power and to become committed. Do not tell them what to do or specify results to accomplish. Simply expect management of the operating units to pick a pilot or pilots and to experiment with them. The general education should emphasize and suggest employee involvement.

Encourage managers to utilize operators on the pilots but do not make it mandatory in the beginning. Select pilot leaders who are not only movers and shakers but who are also inclined to involve people in the right ways.

Objectives for the pilots must be:

- Within the group's span of control.
- Implemented fairly quickly (three months maximum).
- Applied to most of the team members.
- Measurable/tangible.
- Something the group can implement.
- Visible to operators; something that helps them with their job.
- Important to management.

Most of the pilots selected using these criteria will be successful. Use successes with the pilots to educate, develop further commitment, and motivate others to act.

Stages of Manufacturing Excellence Evolution/Process

	Stage	Typical Attitude/Activities
Pre-ME/Traditional	I	"We can't do any better."
	II	"We can do a little better."
	III	"We should do a little better each year": there's some interest in JIT.
Developing commitment	IV	"Let's try pilot(s) and start education." (Employee involvement starts with pilots.)
	V	"The pilot(s) has/have been successful; let's do more."
Breakthrough/ enlightenment (A "critical mass" of the leadership becomes committed.)	VI	"Let's become a world-class manufacturer; manufacturing excellence can be an enormous competitive advantage; it will enable us to survive." Usually the operation forms an operationwide steering committee to oversee the change.
Degrees of implementation	VII	Manufacturing excellence is integrated into the comprehensive strategy/operations plan, budget and objectives, and reward systems. Leadership is proactively involved with building more commitment and momentum.
	VIII	Implementation on one product line.
	IX	Implementation on more than one product line.
	X	The organization has become a world-class competitor throughout all product lines.

The approach to the next phase depends upon the pilots selected and the degree of commitment that exists with management of the plant or operational unit. Typically each plant will want *at least* to do more pilots. Plant leaders can become zealots at this stage and may want to lay out the entire plant. Another consideration at this stage is the degree of commitment to employee involvement (EI). Ideally, the initial pilots will have involved operators. If they did, other operators probably will volunteer to be involved with additional pilots. Operators spread the word when they really do get involved as problem solvers and implementers. On the other hand, if the pilots used a traditional approach to applying JIT/TQC principles (i.e., management, supervision, and indirect involvement only), then it may be necessary to specify that the next pilots involve direct and nonsupervisory employees. In the second round of pilots, expect EI teams to be formed.

EI is probably the most difficult cultural change to make because management and supervisory roles change significantly, which can be very threatening. Instead of making all of the decisions related to process improvement and telling people what to do, managers and supervisors become facilitators of a process that allows the creative energies of all persons, including nonsupervisory personnel, to be utilized to the fullest. Supervisors and indirect staff become a resource for the line in this endeavor. Their roles become supportive as planners and coordinators. Most nonsupervisory personnel enjoy the opportunity to get involved, making improvements that management believes are important. After pilots using the EI approach are successful, just ask for volunteers for the next round of problem-solving teams; there will be more than enough.

Purpose of EI Teams. When getting started, employee involvement teams serve several purposes. They include:

- Helping develop capability.
  Improving leadership.
  Increasing sense of mission and trust.
  Training for leaders, facilitators, and trainers of subsequent groups.
  Improving individual capability.
- Improving productivity and quality utilizing manufacturing excellence principles.
- Helping people better understand principles.
- Helping instill the principles into the culture.

Do *not* expect a certain percentage of improvement at this time if management is sufficiently committed to implement EI. Unfortunately, in some operations significant operational improvements may be required to develop sufficient commitment to proceed further.

FIGURE 7-1
Head Count Required

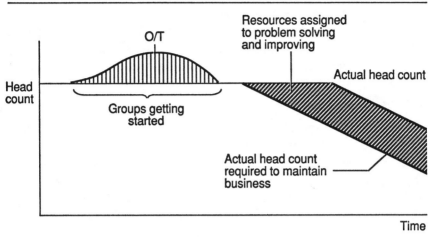

Emphasize development of individual and team capability. Reward success that is consistent with the desired operational improvements. Teams will naturally work on those activities that are rewarded.

Improvements will come out naturally as a result of the improved capability of the group to problem solve. If only results are emphasized initially in a traditional environment, then results are achieved but not improved EI capability.

The long-range goal is to involve nearly 100 percent[2] of *all* personnel as problem solvers/implementors an average of 10 percent of their time. Over a period of time the head count savings from the various teams is reinvested in the organization by allowing additional people to problem solve. Overtime is usually required in the beginning and for a time thereafter to provide problem-solving training and to allow the teams sufficient time to learn how to use new problem-solving skills.

Figure 7-1 depicts head count requirements over time as more people become involved as problem solvers. It assumes constant product demand.

As head count required to do the ongoing work is reduced, the

---

[2]Some good people probably will not ever want to be involved in teams.

actual head count is not trimmed until all persons are involved as problem solvers an average of 10 percent of the time. This may occur three to five years after initial efforts.

To prepare for the time when head count will be reduced, satisfy increased head count requirements due to increased demand by using temporary help and subcontracting to provide a buffer. This buffer plus attrition will reduce the impact on the permanent core of employees.

Change facilitators (i.e., change agents). Change facilitators must be assigned to 1) Help the overall change process at a plant, operation, division, group, and/or company, and 2) Help the problem-solving groups and their participants to accomplish their tasks and to develop individual and collective capabilities. Facilitators of the overall change process help management become committed to ensuring:

- *That all levels and functions within the organization receive a constant educational exposure of some kind* (e.g., managers meet with managers from other plants implementing manufacturing excellence to share what they have learned and plan visits to other companies with JIT/TQC/EI in place, Association for Manufacturing Excellence workshops, and the like). Ensuring understanding throughout the operation is especially critical when the teams begin to bring support groups into their problem-solving process. The support groups need to be "on board" with the concepts or there could be detrimental conflict. Directors need to be on board to establish the right expectation among their managers and department heads to ensure adequate involvement and cooperation by their people.
- *An ongoing plantwide communication* of what is going on, why, who is involved, when the uninvolved might get a chance to be involved.
- *Positive reinforcement* by the directors and managers in the plant where the action is happening. This takes more of the directors' and managers' time, which requires greater commitment.
- *Unity of purpose throughout the organization* relative to individual performance measurements.
- *All organizational levels and functions are engaged* (i.e., involved) appropriately in the overall change process.

When a "critical mass" of operations managers become committed to doing these things, facilitators of the overall change process will not be required.

Facilitators of specific teams. Facilitators of specific teams help teams to process problems correctly and do the following:

- Assess the change process: determine how well groups are accomplishing their purpose and what is required (i.e., leadership, knowledge, will/energy).
- Coach the leader and provide psychological support during role changes.
- Help the leader to do his/her job more easily.
- Provide feedback to the leader and participants.
- Show by example how to facilitate.

Education and training. Implementing manufacturing excellence requires a commitment to continuous education and training. As stated previously, an average of 10 percent of the time of those persons involved on problem-solving teams must be allowed for problem-solving activities. During the first two to three months of team activity, additional time is required for formal education and training.

The education of team members should provide the minimum required to get started—it is important to get into the "learn by doing" mode as soon as possible. Provide only as much education as can be applied immediately; otherwise, it will soon be forgotten. Unfortunately, too many companies get carried away initially and put people into classrooms for one month before they get started. Most people are not used to this and get bored. Rarely are they able to apply everything they have been exposed to, so it is best to spread out the training.

The recommended dosage for team members for the first two months follows. Team members include the leader, facilitator, and participants.

When the teams are well under way (after the third month), education and training can be provided on more of an ad hoc basis. For example, it may be appropriate to provide the following educational sessions:

- Routine maintenance.
- Advanced statistical techniques.
- Team building refresher.
- Specific technical education about an operational process team members are investigating.

The time for education should come from the overall average of 10 percent suggested for problem solving and process improvement. It should probably average 3 to 5 percent.

In the long term, "continuous learning" must be institutionalized and exist in the following forms:

1. Learning objectives are a part of an individual's performance appraisal.
2. Train-the-trainer programs give experts in the organization the skills to design and deliver courses that would otherwise require professional educators.
3. EI team members train and are trained by one another. The team designates "experts" of particular processes.

	Hours
Provide the following before the team starts:	
• Manufacturing excellence overview.	8
Why we are doing this explained by a general manager.	
What it is.	
• Effective employee involvement.	6
What it is.	
Why do it.	
How to do it (includes team building, project management, conducting meetings, and participant roles).	
• Problem-solving basics.	6
• Effective leadership and facilitation for leaders and facilitators.	16
Provide the following after the team has defined the pilot scope and developed a plan to execute the pilot:	
• Quality improvement and statistical process control (provide four hours per week over four weeks).	16
Regardless of whether or not they are involved directly, initially all managers and supervisors should receive:	
• Manufacturing excellence overview (includes discussion of employee involvement).	8
• Quality improvement/problem solving/statistical process control overview.	6

4. Members of work teams receive pay for knowledge as additional skills are learned.

Ideally, employees will think of their organizations as institutions of continuous learning, wherein all employees are involved in training needs analysis through training evaluation.

Where Employee Involvement Can Lead

Essentially two types of EI teams—ad hoc and permanent—exist. The former is established to work on specific tasks. The latter usually is established work units that continue to work on improving the specific environment.

The first leader of the permanent team can be the supervisor of the group. However, as the group develops, it can collectively perform more of the duties supervisors traditionally have done, such as:

1. Interviewing and selecting new group members.
2. Reviewing performance of team members.
3. Disciplining members as appropriate.
4. Communicating and coordinating with other groups and managers.

Over time, the supervisor evolves more and more to a facilitator. Ultimately, groups evolve to where a supervisor is be required. Team members take turns leading team meetings in which the team collectively performs "supervisory" activities. Converting a traditional work unit to a "supervisorless" team typically takes four to five years.

Some permanent groups are established to coordinate the different functions and groups required to design and manufacture a product. Traditionally, different functions have tended to suboptimize their functions at the expense of the whole product. These product-focused groups ensure that a global view is maintained. Ideally, all functions (i.e., marketing, finance, engineering, manufacturing) and physical assets can ultimately be dedicated to a given product or family of products to establish a "business within a business." A product-focused organization is a logical extension of the concept of product-focused cells in manufacturing described in Chapters 2 and 6.

Figure 7–2 compares the traditional organization with a product-focused organization.

FIGURE 7–2

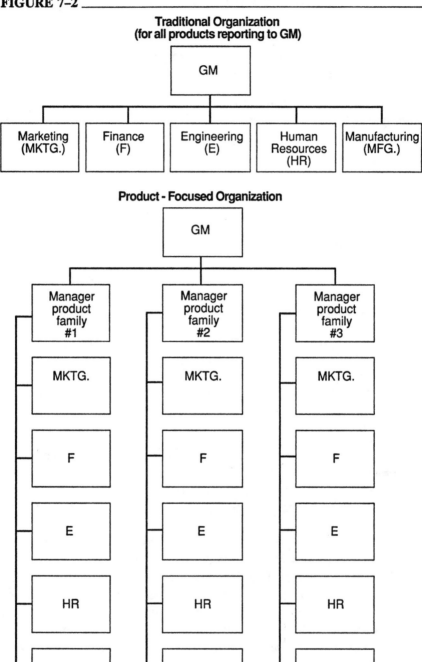

Traditional Organization
(for all products reporting to GM)

GM

| Marketing (MKTG.) | Finance (F) | Engineering (E) | Human Resources (HR) | Manufacturing (MFG.) |

Product - Focused Organization

GM

Manager product family #1 — MKTG. — F — E — HR — MFG.

Manager product family #2 — MKTG. — F — E — HR — MFG.

Manager product family #3 — MKTG. — F — E — HR — MFG.

Clearly, a completely product-focused organization may not be practical because it may require excessive duplication of resources. However, doing it to the furthest degree practical is desirable because it helps:

1. Gear people's thinking to optimizing efforts around the product. By comparison, the traditional organization promotes optimizing a given function frequently to the detriment of the whole.
2. Reduce the total time to design and manufacture new products and to change existing products.
3. People to develop a holistic view.
4. Promote an external orientation to the customers.

Initial reaction to the concept of product focus is that it will require additional people and equipment. In practice, it usually reduces both if the principles of manufacturing excellence are employed in the office as well as in the shop. One manufacturer of custom-engineered valves reduced lead time of order processing, proposal generation, and design from 28 to 4 weeks. As a result, this company's short lead times give it a distinct competitive advantage.

CHAPTER EIGHT

Overcoming Cultural Barriers— Major Pitfalls

Introduction

Many companies get off to a great start with the philosophy of manufacturing excellence. Their success with setup reduction pilots verifies its advocates' assertions. Initial success motivates starting more pilots to further spread the work and develop commitment. Success frequently prompts an across-the-board reduction of safety factors in the MRP system, which reduces WIP by 25 to 33 percent without ill effects. But at some point, progress starts to slow. Statistical process control does not get beyond charting the data. Although originally planned, preventive maintenance does not get going. Decisions do not reflect strong commitment. For example, an automated storage and retrieval system is approved instead of working to eliminate the need to store inventory. A new cell layout is planned, but it has too much space for inventory. Essentially, the improvement process gets stuck. The company may even revert to its traditional ways and lose earlier gains.

At first the slowdown in progress is rationalized as being expected. This is true to a degree, but it is more than that. Things bog down because of cultural impediments; that is, traditional thinking stands in the way of developing a company that improves continuously in all aspects—service, quality, flexibility, productivity. In order to progress beyond the "pilot state," management must be committed to a new way of thinking about manufacturing. This chapter explains how to *free mental logjams, develop commitment, and build lasting momentum.*

MAJOR PITFALLS— WHY ORGANIZATIONS GET STUCK

There Is Insufficient Time to Make Improvements

Many companies find the resources to try a pilot or two, but, to go beyond that requires including improvement activity in the budget, and some companies are not willing to do that. To really involve direct labor people requires budgeting time for them to problem solve and to be trained; this contradicts the traditional approach of maintaining minimal direct labor to support known demand and of allowing only administrative and indirect personnel to work on improving the process. Frequently, when direct labor begins to problem solve, inadequate support is available from the indirect groups. Involving more people than can be responded to is a common pitfall.

Even when companies budget sufficiently for direct labor's education and involvement as problem solvers, some lack the discipline and persistence to execute this budget plan during all phases of the company cycle. When times are very good, everyone must work to get the product out the door. When times are very bad, education, training, and involvement time are cut to retain short-term profitability.

The Plant Tries to Do Too Much

Many plants attempt more improvement projects than can possibly be accomplished. Consequently, efforts are diluted and not much gets done. Improvement activities frequently are not integrated, resulting in duplicating and conflicting efforts. Instead of complementing each other, departments compete with one another. Frequently, projects are suboptimized at the expense of the whole plant. Maintaining focus requires a strong discipline among plant leadership. Looking at JIT as a philosophy of excellence for manufacturing companies and seeing all improvement activities as ways to implement the philosophy help provide integration.

There Is Insufficient Understanding/Education

All too frequently, JIT is still not thought of as part of a philosophy. Instead, it is viewed as an inventory program or, at best, a bunch of improvement techniques. When this happens, JIT is just one of many other improvement activities, and some of these others may get priority at JIT's expense.

A strong tendency exists to give priority to "hard" investments in equipment and tooling over "soft" investments such as education, training, and employee involvement. Direct labor employees are sometimes expected to become problem solvers without adequate training in JIT and TQC techniques, group dynamics, and team building. In fact, the manufacturing excellence philosophy should guide an organization in all endeavors, including the decision-making process for hard investments, such as CIM related items. Subsequently, many capital investments are not required or attempted because the plant does not have the essential prerequisites for automation in place, such as short setup (less than 10 minutes), processes in statistical control (i.e., process variation due only to random causes, not assignable causes), preventive maintenance, and so on.

Some companies naively believe they will receive most world-class benefits when they convert from a "push" to a "pull" system, or that they must convert to "pull" to uncover the "rocks." Actually, converting to a "pull" system should be viewed as "the icing on the cake," to be done after mastering other JIT elements. Companies already know about these "rocks"—managers just need to ask their employees on the shop floor. Most companies can reduce rejects and scrap by two thirds before they implement SPC, if they would just listen to what their employees have been saying for years and really do something about these problems. The point is that a "pull" system is not needed to identify more problems. Listening to employees and then organizing the workplace by removing all inventory, gauges, tools, and other items not required in the immediate time period will uncover more problems than the organization will know what to do with.

Conversion to a "pull" system depends on the case. Generally, it can occur as soon as standard routings do not have regular exceptions and inventory will fit into production space.

Management Control System: The Performance Measurement and Reward System Are Inconsistent

Individual incentive systems are dramatically contradictory to the philosophy because they strongly encourage production, whether or not parts are needed. Rarely are individual's incentives adjusted to reflect production of bad parts. Some companies using this compensation scheme even give credit for performing rework.

Companies still overemphasize direct labor productivity as one of the most important performance measurements. This also encourages production, whether or not it is required. Furthermore, the direct labor productivity measurement is based upon standard hours produced and is rarely adjusted to reflect reworked or scrapped parts. In addition, it results in maintaining direct labor personnel at the lowest possible level, so there is little time for them to become problem solvers.

Holding management accountable for short-term profitability when a full absorption cost accounting system is used also promotes building inventory above needs. It enhances paper profits by absorbing more burden. Also, many companies still do not charge operations for inventory, or when they do, the amount is too small (e.g., 12 percent). This small amount is based upon the companies' perceived cost of capital. In reality, other costs exist such as material handling and control-related expenses, obsolescence, insurance, plus the more obscure costs of masking problems (e.g., scrap, rework, and so on). An inventory cost of 24 percent can be justified easily. If the cost of masking problems is included, the value is probably as high as 48 percent.

Many managers still are rewarded for being "super" expeditors and for making decisions for and commanding their subordinates to do various things. Instead, a manager should be rewarded for creating an environment in which people can function as problem solvers in a more self-directed way. Managers must change from doers and directors to coaches and facilitators of the improvement process. Some companies do not realize this enormous role change is required and become stuck. Managers need significant training and psychological support to help them make this change. Failure to recognize the supervisory role change and the resulting insecurity among supervisors are the primary reasons middle-first-line managers resist the changes more than other management levels. The highest organization levels are insulated from the operating personnel and, therefore, do not feel as threatened.

The Boss Gives It Lip Service

No matter how much is said about a new endeavor, the boss's behavior is what counts. Behavior reflects a person's true commitment and people read the boss's behavior extremely well. Manufacturing excellence requires a cultural change, a change in values and beliefs about how a manufacturing company must be run. Therefore, the boss's behavior must indicate he or she believes in the new philosophy. If the boss expresses "quality first," but compromises quality at the end of the month to make the production schedule, he or she loses all credibility. People will know the boss is not serious, and progress will bog down. In fact, even when the boss's behavior is consistent with the new philosophy, he or she will have to work hard and long to overcome the old beliefs. It takes sustained behavior over time for most people to really believe, making cultural change difficult.

Passive approval to proceed with some pilot projects is not enough. The boss must actively support the evolution and personify the new mentality in all actions. Active support may mean removing people who resist the change. Lew Springer, vice president of manufacturing at Campbell Soup and champion of manufacturing excellence, believes in being patient because the change is so significant. Springer expects that three to six months of continued exposure will be required for people to "get religion." However, he says, "It's like putting everyone in a room and slowly starting a bulldozer the width of the room. Although the bulldozer moves slowly, eventually everybody has to get on board." Without strong leadership, significant progress will not occur. After difficult and stressful change, organizations tend to reach a "comfort level" and will not want to improve further. Leadership needs to keep the pressure on and the momentum going. Creating a culture of continuous improvement is the ultimate accomplishment and requires an enormous amount of sustained emphasis. The first manufacturing excellence operation of any real significance at FMC reverted back to traditional ways after the champion left, even though the philosophy had been evolving for two years!

People Are Not Really Involved

Significant improvement (i.e., becoming a world-class competitor) requires engaging the creative energies of all employees. If all employees are not problem solving, then their creative energies are not being utilized. If a company is serious about employee involvement, 5 percent of everyone's time will be budgeted for training. Employee involvement is the keystone upon which manufacturing excellence is based. Continuous improvement requires problem solving, which in turn requires an environment in which employees can express problems openly, without fear of reprisal. The traditional culture in U.S. factories holds that good managers must have things in control; that is, managers already know about all problems.

The new philosophy requires managers to view their main objective as the unrelenting search for problems. They must create an environment that encourages people to identify problems. Managers must accept that they do not know about all of the problems, and that is OK. This acceptance, in itself, is a huge cultural change.

People who are afraid to bring up problems are also reluctant to try new things because of the fear of failure. This situation makes companies overplan to avoid failure. Little is started. In stark contrast, the "spirit" of manufacturing excellence requires experimentation and accepts that blind alleys and duplication of effort come with the territory.

People Fear for Their Jobs

Unless the company's markets expand after some improvements are implemented, it does not take long for people to understand that widespread application of the philosophy will dramatically reduce head count. Consequently, their willingness to contribute further can stop abruptly. Although a suggestion may not affect the suggestor, it can affect a longtime friend.

It is imperative to share financial and competitive data with all employees, including union members, but this can increase fear. Companies are taking the following measures to deal with the security issue:

- Management issues a policy statement stating that head count will not be reduced due to productivity improvements. Head count can be reduced, however, due to reduced demand and restructuring. A corollary to this policy is a commitment to retraining.
- Whenever possible, increased people requirements are satisfied with temporary employees and/or subcontracted work. If demand is subsequently reduced, temporary and subcontracted work would be reduced first.
- Instead of laying off a given number of employees, everyone

from the top floor to the shop floor would take the same percentage reduction in compensation.

* A significant percentage (i.e., 50 percent) of the company's compensation is based upon how well the company does overall. If the company does poorly, everyone is affected equally and no jobs are lost.

People Believe They Are Not Sharing in the Benefits

In some companies, security is not an issue. Demand is increasing at a faster rate than productivity improvements. In this case, people can easily grow to believe that profitability is being improved through their involvement and hard work, but they are not sharing in the prosperity to the same degree as management. Being recognized for achievement is important but alone is inadequate when people *perceive* that management is not fair with respect to compensation. Having a profit-sharing plan that accounts for a significant percentage of compensation and operates like Lincoln Electric's (described in Chapter 4) prevents this problem.

The Champion Leaves Too Soon

Tragically, in some companies the management champion of a business unit may leave before the new thinking is institutionalized. Changing culture usually takes longer than the tenure of managers, especially those most likely to lead a movement like manufacturing excellence. This is a concern of the top management champions. It is another reason why there must ultimately be a top management champion, because only this level is in a position to deal with this issue.

Overcoming Cultural Barriers—
Building Momentum

After avoiding or overcoming the major pitfalls discussed earlier, manufacturers need to build momentum and create a culture of continuous improvement.

Ensure "Constancy of Purpose"

Constancy of purpose with regard to improvement of product and service is critical; everyone must be marching to the same drummer. W. Edwards Deming placed it first among his 14 points that ensure competitive advantage.

First, make certain a constancy of purpose exists among top managers and their staff. An effective way to do this is to send these people off the site to develop a vision of what the operation will look like after the philosophies and concepts have been in place for three to five years. Ideally, the vision will be expressed in terms of "premises of manufacturing excellence." The appendix gives a suggested list of premises. These principles should address the performance measurement and reward system and should be considered a "living document," one that is reviewed and updated periodically. After developing this vision, a strategy for attaining the vision is developed. The strategy's objective is to make the mind-set of the operation consistent with the premises.

Develop a Realistic and Integrated Plan

The best way for an organization to begin this journey of continuous improvement is to conduct general JIT/TQC education, pick a small pilot or pilots, and get started. Avoid planning the evolution in minute detail. Look at the pilots as experiments and as ways to learn more about the philosophy and principles.

After completing these initial pilots, initiate more, ideally in other areas to promote exposure throughout the organization. Over time, expand the number and scope of the pilots. Eventually the pilot activities may start to duplicate or compete with one another for resources or there may be insufficient resources for more. When any of these situations arise, it is time to develop a realistic and integrated plan for further improvement.

Prepare a "rough cut" capacity plan for all improvement projects in progress and in planning, including those not yet under the manufacturing excellence banner. Estimate the time required for different resource areas (e.g., product engineering, process engineering, direct labor) and compare with available resources, making allowances for turnover and a 10 percent contingency. Delete and adjust projects until the available resources are in sync with required resources. Because this process can become overly complex, allow no more than one month to form a realistic estimate with the rough cut numbers. The general manager's leadership is usually severely tested by this exercise because people often want to do more than available resources allow. The general manager may have to cut a number of pet projects. When led properly, the process not only attests to the general manager's commitment, but also clarifies the rationale for supporting a manufacturing excellence project over a more traditional effort. Make sure the plan is consistent with the measurement and reward system.

Organize to Provide the Right Direction and Support

Establish a steering committee led by the general manager and consisting of other members of the GM's staff. The committee's purpose is to:

- Approve and recommend projects.
- Appoint ad hoc multifunctional teams to work on specific problems and projects.
- Ensure resources are available.
- Review progress.

- Revise priorities as required.
- Applaud success.
- Evaluate its own behavior relative to the JIT premises.

The committee should meet at least every two weeks at the beginning of the change process. All members should have a general understanding of the various techniques being utilized (e.g., statistical process control). Members should be able to intelligently discuss a shop floor control chart with operators.

Recognizing and applauding success is a significant change for most organizations. A structure must be provided to ensure success is acknowledged appropriately and sufficiently until such acknowledgment becomes second nature. The structure may be an ongoing action item on the steering committee's agenda to identify success, to review ways the success was applauded, and to evaluate the committee's efforts at applauding successes.

Because leadership's behavior counts so much when changing culture, the steering committee must evaluate its behavior relative to the premises of manufacturing excellence. Like New York City Mayor Ed Koch, the steering committee needs to continually ask itself and others, "How are we doing? Is our behavior consistent with the new philosophy we espouse?" This feedback from operators and all others involved is essential because it is their *perception,* accurate or not, that counts. Asking people these questions spontaneously on a walk around the plant is a powerful way to break down barriers between management and employees. The questions tell people management wants to change its own behavior too and manufacturing excellence is not just one more thing management is foisting onto them.

Reinforce Positive Results

Positive reinforcement of efforts and achievements consistent with manufacturing excellence is a powerful way to build momentum; it accomplishes the following at the same time:

- Satisfies a basic human need to be considered competent and worthy by work associates.
- Establishes what is expected. What receives most attention, negative or positive, is considered to be most important.

- Educates by providing more exposure to the philosophy and principles.
- Makes the person who is doing the reinforcing feel good.

It is especially important that the leaders provide positive reinforcement for a couple of reasons. First, it usually means a lot to those receiving the recognition. Second, it further develops commitment by the leaders. Getting someone to positively reinforce something is like getting them to say "yes" when selling them something. The more they reinforce it, the more they behave in a way consistent with what they are reinforcing.

Positive reinforcement can be provided in many ways—company publications; name or picture on a bulletin board; applause at a plantwide steering committee; a videotape of what was accomplished sent to other groups or plants as an example of the right way to do things; mention in the performance review; saying, "Thank you, you are doing a great job"; attention to a recommendation and support to do it; acceptance of the next recommendation on faith because of the previous success, and many, many more.

Positive reinforcement is new to most people. Consequently, until it becomes second nature, a structure should be established to ensure it is done sufficiently. Every plantwide steering committee meeting should evaluate how well and frequently it positively reinforces good results, especially relative to negative reinforcement. "Positives" should outweigh "negatives" by 3 to 1. The duration of positive reinforcement should be three times the duration of negative reinforcement.

People want to be on a winning team. A winning spirit is created only by positive reinforcement. Successful sports teams continually reinforce success. There is a high degree of praise and encouragement. Sadly, few companies do this but it is a common trait of great companies. IBM, for example, sets sales quotas so that roughly 80 percent of account executives meet them. IBM has a strong culture of applauding success.

Share Business Results and Competitive Data

Sharing business results and competitive data builds trust and helps people better understand management decisions. In fact, when people understand the business and competitive environment, most come to the same conclusions as management. Tom Rabaut, a division manager of FMC and an outstanding leader, believes in explaining the business realities, letting the people draw conclusions, and listening to what they think the business should do. According to Rabaut, "Being open is a lot easier than being closed and then having to justify actions. How else are all employees going to identify with the business? They must see the business as managers do." Openness builds trust and results in the open identification of problems. This is essential if all people are to become problem solvers and if a firm wishes to derive the maximum benefits of manufacturing excellence.

Rabaut regularly shares the income statement by product line, return on assets, and break-even analysis with all employees. Considerable time was invested to ensure people understood what these measurements meant and how they were derived. When Rabaut's operation was instituting world-class improvement activities, a machinist stopped Rabaut to remind him he was behind on his commitment to sell equipment and this was preventing the group from making its break-even point reduction objective!

In contrast, at a plant of a power generator manufacturer, an operator was observed cutting off six feet of unused material after each cycle of an operation and throwing it into the scrap bin. The manufacturing engineer commented that scrap cost amounted to about $6,000 per year and the operator would not be that wasteful if it were his company. As long as a supervisor was looking over the operator's shoulder, the operator did not waste the material. However, when the number of supervisors was reduced to cut costs, the operator reverted to his old ways. The answer to this situation is not more supervisors but to share the realities of the business so people identify with the company as if it were their own. Every employee touches these $6,000 cost reduction opportunities every day. The spirit of manufacturing excellence creates a unity of purpose among *all* employees so all of these opportunities are realized.

Give People a Sense of Mission

Although employees need to understand business results in terms of profit and loss and return on assets, they need even more to be committed to becoming a world-class competitor. This is what the philosophy is all about. Management and other personnel need to have a sense of mission, of cause, and of purpose; that is, a sense that their work is doing more than just making profits for the company (and themselves, if they are stockholders) and more than providing them with a good source of income. Following are examples of mission, cause, and purpose:

- *Survival.* This has been the mission of most U.S. companies that have invested significantly in world-class manufacturing.
- *Being best in the world at what they do.*
- *Patriotism.* The great Japanese companies have used this as a powerful mission; they have motivated people to commit long hours to becoming competitive worldwide for the benefit of the whole Japanese economy.

"Survival" may be the right mission for your company if markets are eroding due to intense competition. But if return on net assets exceeds 25 percent, and market share is secure, emphasizing the "being the best" theme would be better. All three themes are valid if a long-term perspective is maintained. When viewed over the long term, all businesses are in a survival mode; any differences among companies are only a matter of degree. Extremely profitable markets, those with a return on net assets that exceeds 25 percent, are a prime target because of their high returns.

Attaining the philosophy's benefits requires a high degree of commitment from all employees that comes only from a powerful sense of mission. Leaders must espouse more than a philosophy; they must espouse mission.

Stressing survival when economic conditions are good is difficult but can and must be done. Lew Springer of Campbell Soup is a good example. He has continually talked about long-term survival and how the Japanese would eventually enter the U.S. food processing markets and start competing with Campbell Soup. Campbell has always been profitable. In January 1985, Springer expressed an urgent concern about implementing manufacturing excellence philosophy because he expected the Japanese to make their move soon. In April of that same year, a California newspaper reported a Japanese firm had just purchased a small U.S. soup company in the Fresno area.

Summary

Implementing the world-class manufacturing philosophy, which is based upon continuous improvement, is much more than implementing a set of manufacturing techniques. It is a philosophy of excellence for manufacturing companies that requires a new way of thinking. In short, it is a cultural change. Cultural changes are very difficult and require an enormous degree of will, energy, and leadership. Cultural change must be led by management. In fact, the top manager's behavior is the most critical element. If his or her behavior is perceived to be consistent with the new philosophy, change will occur. Even when the boss's behavior is consistent, it needs to be sustained for a long time for most people to believe the change is real and permanent.

It took Toyota 25 years to attain its present state of excellence. After the oil embargo hit Japan in 1973, other Japanese companies, whose mission was survival, did it much faster (e.g., Toyo Kogyo).

Omark Industries, a U.S. company purchased by Blount Construction and a market leader when it started, will probably reach a world-class level in 1987. Omark's journey was led by Jack Warne, chief operating officer. Chrysler Corporation in Australia has nearly attained this level in only two years as a result of a severe survival mode.

Motivated by an intense sense of patriotism, a Detroit automobile plant was converted to an aircraft plant during World War II. After the decision to convert to an aircraft plant, the first plane rolled onto the runway in one year to the day. When Americans are motivated by a strong sense of mission, the United States can become best in the world. The present situation is precarious for U.S. manufacturing companies. A strong sense of mission is essential in developing the intense degree of commitment required to make the JIT philosophy a lasting reality.

The Spirit of Manufacturing Excellence

In many ways, looking first at implementation and strategy is an inverted approach to achieving manufacturing excellence. New techniques or structures succeed because they are management's expression of an accepted value or spirit. Of course, the answer to improved productivity will not be found simply in techniques. It will be found in acceptance of and action on the values and attitudes that gave birth to the present world-class techniques.

Acceptance of new values and spirit has the power to create new cultures of excellence. These motivating values create a willingness to sacrifice personal gain for a higher vision. The American government did not emerge because a technician perceived that a certain set of checks and balances were superior to other forms of government. On the contrary, a new ideal was articulated in the Declaration of Independence. This ideal gave birth to a radically new political organization and the spirit of our nation.

Values and a spirit are at the foundation of how we organize our corporations. Today, the financial necessity of improved productivity and quality demands a new corporate philosophy and culture. As the introduction indicated, American products are being challenged everywhere and the U.S. standard of living has even begun to decline. To meet the challenge of global competition, corporations must fulfill their employees' deep need for meaning and significance in their life work.

Such a sense of meaning and higher purpose is the foundation of motivation and, therefore, of productivity and economic survival.

Unfortunately, business largely fails to express its purpose in terms that inspire its people.

America is entering a new era: global, interdependent, intercompetitive. As a result, management's role has radically changed. Management must make the leap from an adversarial and combative attitude to one of cooperation and trust. Several new facts of doing business support this. The nature of work is more cognitive. Even factory workers are now "knowledge" workers. Technological innovations arise from all levels of the company. Managers today must cause people to think creatively, to learn, and to share. Second, individuals have many more options. Management through intimidation is dead; management through employee involvement and positive reinforcement is required. Third, motivation by material reward alone is increasingly inadequate. Successful managers will inspire a higher purpose in their subordinates. A shift to committed cooperation will reduce the need for managers. A properly encultured employee will be involved, committed, and rewarded so as to ensure his or her performance. The new manager's purpose will be to create and sustain this commitment. In global competition, managerial competence is a critical competitive advantage.

The most difficult part of this transition is close to home. Developing specific skills such as listening, problem solving, and reinforcing is not difficult. The difficult task is using them if they go against a manager's fundamental beliefs about what a manager is. Managers are largely unaware they are operating in blind obedience to the conditioning of years of movies and television that proclaimed war or western models of how to achieve success and victory. An alternative set of beliefs and models about how management manages must be adopted by all levels, but especially higher management.

The distinction between the manager of the past and the new manager is the difference between a boss and a leader. It can be summarized by the word *purpose*. Leaders have a noble vision of their purpose. Leaders create energy and enthusiasm by instilling purpose in others. Bosses only control and direct already existing energy. This stifles innovation. Leadership brings out the creativity of the individual by showing that the success of a product or service is a genuine contribution to the good of all (i.e., a higher purpose).

The new managers are transformational leaders. They strive to inspire purpose because it is the seed of motivation. Understanding that an enterprise has a higher purpose makes employees willing to sacrifice, and such sacrifice enhances their own dignity and self-esteem. Honeywell's Aerospace Group worked long hours with total dedication to achieve the noble purpose of placing an American on the moon. Many major corporations emerged because of the spirit of their founders. Henry Ford's deep commitment to meeting the need for cheap, dependable transportation for all Americans sustained Ford, as well as those who followed him in pursuit of his vision.

To maximize return on investment is an inadequate statement of purpose. It fails to accomplish the paramount responsibility of leadership—to provide meaning and inspiration to those who are expected to follow. The mass of people within our corporations are not primarily motivated by maximizing private gain. Many firms are structured as if they were. Thus, it is not the norm to be highly dedicated and energized by work. In reality, the appeal to spirit, self-esteem, and higher value is the prime mover.

Financial objectives also are an inadequate statement of purpose because they fail to recognize the social justification for the corporation's existence. Business enterprises do have a noble purpose, and they should recognize and proclaim it. The purpose of business is the creation of wealth—not for a few, but for all. The creation of genuine wealth in the form of goods and services must be the corporation's primary purpose. Financial results will follow as the corporation succeeds in this primary aim.

Recently, devotion to a product or service has become less prevalent than devotion to success through financial means such as mergers, tax, or accounting strategies. But such approaches create no new wealth. Corporate executives must realize that devotion to creation of wealth in the form of goods and services will lead to success, while attention to only the organization or consolidation of wealth through financial strategy cannot.

The new manager must be able to inspire his subordinates for another equally important reason. Innovation, the crucial source of today's competitive advantage, comes from workers just as much as from managers. Workers who spend all day on a task usually know how to improve it better than anyone. No manager in a position of significant responsibility can possibly have all the knowledge and experience needed to make decisions. He needs the input of all his subordinates. The shift from a command to a consensus or cooperative leadership is demanded by the material realities of our work.

The authoritarian, commanding manager of the past will be the dinosaur of tomorrow. But to change from a command mode to a cooperative, consensus mode requires surgery on the personalities of many managers. It can and must happen, but it isn't easy. Upper management must accept its responsibility as the exemplary role models of the new philosophy. Subordinates scrutinize their boss's attitude very closely. The most significant element in the success of employee involvement is the degree to which senior managers practice the same methods. If cooperative problem solving becomes the norm, it will last and be accepted by the organization. Such decisions represent the sum of all the knowledge and experience of the group. They are the product of frank and honest discussion of all parties, to which all are willing to commit themselves as if the decision were their own. Better, more creative decisions result, and more determined actions follow.

In the process of inspiring a sense of higher purpose and cooperation, the manager can raise examples of excellence to encourage subordinates to improve their performances. Those achieving excellence are participants in an ongoing struggle with their own competencies. They are never satisfied with their current record but realize there are higher levels of achievement to pursue. A manager can foster such a spirit by presenting examples of excellence, heroes of the craft. Motivation is sustained by setting specific objectives, measuring performance, and rewarding work well done. Absolutely frank, but nonjudgmental, feedback keeps the employee in tune with his efforts.

A spirit of oneness, mutuality of interest, and of concerted action must replace the class and adversarial divisions that now definitively characterize our organizations. A competency continuum that recognizes every employee (from the factory worker to the chief executive) must participate in management, and must perform productive work, should replace the current divisive management-worker system. Maintaining this divisive system is the greatest single barrier to productivity, especially in the manufacturing sector.

False assumptions about labor are inordinately expensive. They prevent utilizing employees' abilities and creative energies. They justify an excessive number of managers, deny the employee the opportunity to accept responsibility for his work, add to cost, and reduce performance. All management practices should be aimed at creating a spirit of unity between people at all levels. The worker is capable of self-management; his being assigned the role of nonthinking, nonparticipating labor is the seed of labor discontent and management inefficiency.

The unity between the individual and the organization can be understood in terms of identification, the state in which the individual believes his own self and well-being is tied to that of the firm. Identification causes the individual to be willing to sacrifice for the good of the whole and to experience pleasure if the firm succeeds.

As American business matures in its understanding of the role of the individual; his needs; the involvement, respect, and self-esteem so critical to his creative performance, a new commitment to caring and community will develop. This fulfillment will become necessary as workers' options increase and as work becomes increasingly cerebral and creative.

Increasingly, it is good business to care. As innovative technology and the rate of its replacement accelerates, the human beings who are the source of innovation become considerably more important to the modern firm, whose competitive advantage relies on innovation. And innovations come just as often from the factory floor as from the corporate office.

A caring culture maximizes production of the human capital in which it has invested. Corporations today succeed because of the creativity of their people. Creativity is linked to the existence of genuine caring and community. It is fostered when small groups of similarly talented individuals fully exchange ideas in the total absence of judgment or fear of condemnation. Innovation and creativity are the lifeblood of a firm's continued existence and future profits. For this reason, a genuine community of caring is very good for business.

There can be no leadership without integrity. Webster's Dictionary defines integrity as "a firm adherence to a code of moral values." It is the quality of incorruptibility, of being complete and undivided. Managers with integrity do what they say they will do and receive the trust and faith of their staff. Staff members gain a sense of security because they feel their environment is predictable and dependable.

Inasmuch as self-management is becoming the dominant theme of excellent organizations, integrity needs to be the characteristic trait of all workers. Self-management can be built only on the trust a firm's integrity makes possible.

When a firm has integrity, its workers believe increased performance, or an innovation on their part, will be properly rewarded and not lead to their being made obsolete and laid off. Employees respond with their best efforts when they sense management is acting in their best interests. It is management's ability to act with integrity, and to be viewed by subordinates as possessing integrity, that permits the movement toward greater self-management and efficiency.

A cycle of trusting and trustworthy behavior is established by the manager who has the integrity and self-confidence to trust others. We intuitively trust the person who places trust in us. Because we perceive the other has high expectations of us, we act to fulfill those expectations.

Managers of high integrity often have a sense of a higher purpose even when their firms are groping for one. Such managers believe in a cause, perhaps a religious or spiritual purpose beyond their control, to which they and their world should conform. Their beliefs provide an internal security that allows them to act with integrity. They are not merely serving themselves, but they are serving a mission for which great sacrifices are justified.

Employees can sense a manager who has a spiritual purpose and respond in ways that add integrity and meaning to their own lives. They, too, will sacrifice for it. When we sacrifice for a noble cause, we find there has been no sacrifice. We become more worthy having made the sacrifice.

A cause or purpose need not be glamorous or remarkable to be involving. It need only represent a genuine desire to better the human lot. Thus, it shares a collective goal with our society's other businesses. It is business and industry that produce wealth, that will eliminate poverty, disease, and ultimately war, and that will free humanity from the chains of mindless toil so we all can develop our higher capacities of mind and soul.

Premises of Manufacturing Excellence

Preamble: Manufacturing excellence is based upon the notions of continuous improvement, simplicity, and elimination of waste. Waste is anything that does not add value to the product or service.

1. Quality is viewed as satisfying the customers' needs. This is the primary orientation of all employees.
2. There is a prevailing commitment to reducing process variation. Controlling the process is the way to ensure product quality. Inspection is at the source. Inspection, production processes, and the product are designed to be foolproof to every degree possible.
3. All employees are expected to be problem solvers; that is: (a) identify problems and alternative solutions, (b) participate in implementation of improvements, (c) participate in establishment of objectives. There is a commitment to developing people and to utilizing their full potential.
4. All persons are informed regularly about their customers' needs, how well their customers' needs are satisfied, and their competitors' approach to satisfying customers' needs.
5. All workplaces (including the office) are well organized; there is a place for everything and everything is in its place, clean and ready for use.
6. Setup reduction is an important objective because it increases flexibility, reduces lead time, and enables very small lot production, which ensures that quality problems are detected quickly.

7. Cellular processes with product focus are utilized to the maximum degree in the office as well as the shop.
8. Inventory is aggressively reduced to unmask problems.
9. Preventive maintenance involving all employees is rigorously employed and is viewed as a way to increase understanding of the process.
10. To the maximum practical degree, administrative and shop flows are balanced and uniform. Ideally, a day's requirements of everything are done every day. The goal is to match daily output with daily customer demand.
11. Products are designed with productibility, the environment, and safety in mind.
12. Plant and office layouts must be as compact as possible to ensure low inventory, which promotes productivity and quality and to facilitate control by visibility.
13. To every practical degree, a highly integrated network will be developed with suppliers and customers. Suppliers are viewed and view themselves as partners in the business. A mutual long-term commitment exists between suppliers and customers.
14. An ultimate objective is to do everything to order. To this end, manufacturing control systems must evolve from complex "push" to simpler "pull" systems to the furthest degree possible.
15. Management's reward is based largely upon the rate of improvement and the degree to which they contribute to employee involvement. The number of ideas generated by all employees and the percentage implemented is an important measure in this regard.

ABEGGLEN, JAMES C., and GEORGE KAISHA STALK. *The Japanese Corporation.* New York: Basic Books, Inc., 1985.

ABERNATHY, WILLIAM J., KIM B. CLARK, and ALAN M. KANTROW. *Industrial Renaissance: Producing a Competitive Future for America.* New York: Basic Books, Inc., 1983.

CROSBY, PHILIP B. *Quality Is Free: The Art of Making Quality Certain.* New York: McGraw-Hill, 1979.

DEMING, W. EDWARDS. *Quality Productivity and Competitive Position.* Cambridge, Mass.: Center for Advanced Engineering, Massachusetts Institute of Technology, 1982.

HALL, ROBERT W. *Zero Inventories.* Homewood, Ill.: Dow Jones-Irwin, 1983.

HAYES, ROBERT H., and STEVEN C. WHEELWRIGHT. *Restoring Our Competitive Edge: Competing through Manufacturing.* New York: John Wiley & Sons, 1984.

HUGE, ERNEST C. "Managing Manufacturing Lead Times." *Harvard Business Review,* September 1979, pp. 116–123.

ISHIKAWA, KAORU. *What Is Total Quality Control? The Japanese Way.* Translated by David Lu. Englewood Cliffs, N.J.: Prentice-Hall, 1985.

JURAN, JOSEPH M., and FRANK M. GRYNA, JR. *Quality Planning and Analysis.* New York: McGraw-Hill, 1980.

MILLER, LAWRENCE M. *American Spirit: Visions of a New Corporate Culture.* New York: Warner Books, 1984.

PETERS, THOMAS J., and ROBERT H. WATERMAN, JR. *In Search of Excellence.* New York: Harper & Row, 1982.

PRESIDENTS' COMMISSION ON INDUSTRIAL COMPETITIVENESS. *Global Competition, The New Reality.* vols. I, II, III. Washington, D.C.: U.S. Government Printing Office, 1985.

SCHONBERGER, RICHARD J. *Japanese Manufacturing Techniques: Nine Hidden Lessons in Simplicity.* New York: Free Press, 1982.

_____. *World Class Manufacturing. The Lessons of Simplicity Applied.* New York: Basic Books, 1986.

SHINGO, SHIGEO. *A Revolution in Manufacturing the SMED System.* Stamford, Conn.: Productivity Press, 1985.

_____. *Zero Quality Control: Source Inspection and the Poka-yoke System.* Stamford, Conn.: Productivity Press, 1986.

SKINNER, WICKHAM. *Manufacturing in the Corporate Strategy.* New York: John Wiley & Sons, 1978.

_____. *Manufacturing. The Ultimate Competitive Advantage.* New York: John Wiley & Sons, 1985.

INDEX

Asset utilization, 66
Automated guided vehicles, 67–68
Automated process planning, 35
Automated storage and retrieval systems
   (ASRS), 67

Balanced operations, 25
Blanket purchase orders, 47

Cause and effect diagrams, 17–19
Cellular layout; *see also* Flexible
   manufacturing systems
   (FMS)
  automated, 32–33
  defined, 23
  implementation of, 64
  product-focused, 23–24
  replication in, 65
Change facilitators
  role of, 86
  for specific teams, 87
  training for, 87–89
Compensation
  based on seniority, 44
  through bonuses, 44
  and employment stability, 44
  through group incentives, 44, 89
  profit-sharing plans for, 45–46, 98
  reduction in, 97–98
Competitive strategy
  in CIM implementation, 63
  and corporate philosophy, 1–3, 44, 46,
    103–4
Computer aided design (CAD)
  defined, 34
  proper use of, 36

Computer integrated manufacturing
  (CIM)
  benefits of, 62–63
  elements of, 34–35
  environment for, 36, 58–59
  risks in implementation of, 62
  strategic approaches to, 59–64
Computer numerically controlled (CNC)
  machines
  as machining centers, 65
  use of, 33
Conformance, quality of
  defined, 15, 52n
  management approaches to, 21
  and statistical process control, 16–20
  as a strategic focus, 52
Control charts, 18
Coordinate measuring machine (CMM), 33
Corporate culture
  change strategy for, 81, 101–4
  and the competitive edge, 103–4
  desirable, 2–3, 30–31, 106–11
  and quality control circles, 13–14
  for total quality control, 3–31
Cost
  of implementing CIM, 61
  of quality (COQ), 43, 46
  reduction in, 55–56
  as a strategic focus, 51
Cox, Allan, 78
Customer service
  as a performance measurement, 46
  as a strategic focus, 14, 43, 51

Demand pattern, 56
Deming, W. Edwards, 5–6, 18, 30, 99
Deming Circle, 18, 20
Distributed numerical control, 35

116

Technology—*Cont.*
  custom-designed, 65
  and depreciated equipment, 66
  equipment selection, 67
  justification of, 69–73
  off-the-shelf, 64–65
  selection of, 64–68
Total quality control (TQC)
  corporate culture required for, 30–31
  and customer needs, 14, 30
  defined, 14, 30
  feedback in, 19–20
  and JIT, 30
  management approaches to, 21, 78
  and process variables, 15
  and quality of conformance, 15, 30

Total quality control—*Cont.*
  and quality of design, 15
Tribus, Myron, 79

Vertical integration, 56

Warne, Jack, 105
Waste; *see also* Inventory
  defined, 12, 67
  elimination of, 6–7, 25–27
Workplace organization
  and cellular layout, 23–24
  to eliminate waste, 23
  shop floor control, 25